T0251008

B-Lymphocyte Differentiation

Editor

John C. Cambier, Ph.D.

Member
Department of Medicine/Division of Basic Immunology
National Jewish Center for Immunology and Respiratory Medicine
Denver, Colorado

CRC Press
Taylor & Francis Group
Boca Raton London New York

CRC Press is an imprint of the
Taylor & Francis Group, an **informa** business

First published 1986 by CRC Press
Taylor & Francis Group
6000 Broken Sound Parkway NW, Suite 300
Boca Raton, FL 33487-2742

Reissued 2018 by CRC Press

© 1986 by CRC Press, Inc.
CRC Press is an imprint of Taylor & Francis Group, an Informa business

No claim to original U.S. Government works

This book contains information obtained from authentic and highly regarded sources. Reasonable efforts have been made to publish reliable data and information, but the author and publisher cannot assume responsibility for the validity of all materials or the consequences of their use. The authors and publishers have attempted to trace the copyright holders of all material reproduced in this publication and apologize to copyright holders if permission to publish in this form has not been obtained. If any copyright material has not been acknowledged please write and let us know so we may rectify in any future reprint.

Except as permitted under U.S. Copyright Law, no part of this book may be reprinted, reproduced, transmitted, or utilized in any form by any electronic, mechanical, or other means, now known or hereafter invented, including photocopying, microfilming, and recording, or in any information storage or retrieval system, without written permission from the publishers.

For permission to photocopy or use material electronically from this work, please access www.copyright.com (http://www.copyright.com/) or contact the Copyright Clearance Center, Inc. (CCC), 222 Rosewood Drive, Danvers, MA 01923, 978-750-8400. CCC is a not-for-profit organization that provides licenses and registration for a variety of users. For organizations that have been granted a photocopy license by the CCC, a separate system of payment has been arranged.

Trademark Notice: Product or corporate names may be trademarks or registered trademarks, and are used only for identification and explanation without intent to infringe.

Library of Congress Cataloging in Publication Data
Main entry under title:

B-lymphocyte differentiation.

    Includes bibliography and index.
    1. B lymphocytes. 2. Cell differentiation.   I. Cambier,
    John C., 1948-   .
    QR185.8.L9B19   1986   616.07'9   85-12726
    ISBN-0-8493-5 172-3

A Library of Congress record exists under LC control number: 85012726

Publisher's Note
The publisher has gone to great lengths to ensure the quality of this reprint but points out that some imperfections in the original copies may be apparent.

Disclaimer
The publisher has made every effort to trace copyright holders and welcomes correspondence from those they have been unable to contact.

ISBN 13: 978-1-315-89091-3 (hbk)
ISBN 13: 978-1-351-07001-0 (ebk)

Visit the Taylor & Francis Web site at http://www.taylorandfrancis.com and the
CRC Press Web site at http://www.crcpress.com

THE EDITOR

John C. Cambier, Ph.D., is Member, Department of Medicine/Division of Basic Immunology, National Jewish Center for Immunology and Respiratory Medicine and is Associate Professor, Department of Microbiology and Immunology, University of Colorado Health Sciences Center, Denver, Colorado.

As a doctoral student in 1974, Dr. Cambier served as a Fellow at the Max-Planck-Institute fur Tierzucht und Tierernahrung, Mariensee, West Germany assisting in the establishment of a research program designed to examine the importance of breast feeding in the protection of infants against diarrheal diseases. He received his Ph.D. in 1975 from the University of Iowa in Immunology-Virology. From 1975 to 1978 he served as an NIH postdoctoral fellow and then as an instructor at the University of Texas Health Science Center. Dr. Cambier held various positions at Duke University Medical Center from 1978 to 1983, including Adjunct Associate Professor in the Department of Microbiology and Immunology. During that time he was a Visiting Scientist at Bundesforschungsanstalt fur Tierzucht und Tierverhalten, Neustadt, West Germany. He has been on the staff of the National Jewish Center for Immunology and Respiratory Medicine and the University of Colorado since 1983.

Dr. Cambier has been the recipient of the National Research Service and the Research Career Development awards and is a member of the American Association of Immunologists, the Flow Cytometry Society, and the NSF Cell Physiology Panel.

He has authored or co-authored more than 50 scientific publications. His major research interests include lymphocyte activation and regulation, stem cell regulation, and flow cytometry.

CONTRIBUTORS

Larry W. Arnold, Ph.D.
Research Assistant Professor
Department of Microbiology and
  Immunology
University of North Carolina
Chapel Hill, North Carolina

Gail A. Bishop, Ph.D.
Postdoctoral Fellow
Department of Microbiology and
  Immunology
University of North Carolina
Chapel Hill, North Carolina

John C. Cambier, Ph.D.
Member
Department of Medicine/Division of
  Basic Immunology
National Jewish Center for
  Immunology and Respiratory
  Medicine
Associate Professor of Microbiology
  and Immunology
University of Colorado
Denver, Colorado

K. Mark Coggeshall, Ph.D.
Department of Pharmacology
Washington University School of
  Medicine
St. Louis, Missouri

Ronald B. Corley, Ph.D.
Associate Professor
Department of Microbiology and
  Immunology
Duke University Medical Center
Durham, North Carolina

Thomas L. Feldbush, Ph.D.
Professor
Departments of Microbiology and
  Urology
University of Iowa
Veterans Administration Medical
  Center
Iowa City, Iowa

Fred Finkelman, M.D.
Director
Division of Rheumatology and
  Immunology
Associate Professor
Department of Medicine
Uniformed Services University of the
  Health Sciences
Bethesda, Maryland

Geoffrey Haughton, Ph.D.
Professor of Immunology
Department of Microbiology and
  Immunology
University of North Carolina
Chapel Hill, North Carolina

John W. Kappler, Ph.D.
Associate Professor
Department of Medicine
National Jewish Center for
  Immunology and Respiratory
  Medicine
Denver, Colorado

David E. Lafrenz, Ph.D.
Research Scientist
Research & Development
Veterans Administration Medical
  Center
Iowa City, Iowa

H. James Leibson
Department of Medicine
National Jewish Center for
  Immunology and Respiratory
  Medicine
Denver, Colorado

Nicola J. LoCascio, Ph.D.
Research Assistant
Department of Microbiology and
  Immunology
University of North Carolina
Chapel Hill, North Carolina

Paula M. Lutz, Ph.D.
Department of Microbiology and
  Immunology
University of North Carolina
Chapel Hill, North Carolina

Phillippa Marrack, Ph.D.
Associate Professor
Department of Medicine
National Jewish Center for
  Immunology and Respiratory
  Medicine
Denver, Colorado

Eleanor S. Metcalf, Ph.D.
Associate Professor
Department of Microbiology
Uniformed Services University of the
  Health Sciences
Bethesda, Maryland

James J. Mond, Ph.D., M.D.
Associate Professor
Department of Medicine
Uniformed Services University of the
  Health Sciences
Bethesda, Maryland

John G. Monroe
Department of Medicine
Tufts New England Medical Center
Boston, Massachusetts

Professor Sir Gustav J. V. Nossal,
  F.A.A., F.R.S.
Director
The Walter and Eliza Hall Institute of
  Medical Research
Victoria, Australia

Christopher A. Pennell, B.S.
Research Assistant
Department of Microbiology and
  Immunology
University of North Carolina
Chapel Hill, North Carolina

Beverley L. Pike, Ph.D.
Cellular Immunology Unit
The Walter and Eliza Hall Institute of
  Medical Research
Victoria, Australia

John T. Ransom
National Jewish Center for
  Immunology and Respiratory
  Medicine
Denver, Colorado

Neal Roehm, Ph.D.
Research Associate
Department of Medicine
National Jewish Center for
  Immunology and Respiratory
  Medicine
Denver, Colorado

Laura L. Stunz, Ph.D.
Graduate Student
Department of Microbiology
University of Iowa
Iowa City, Iowa

Dorothy Yuan, Ph.D.
Assistant Professor
Department of Pathology
University of Texas Health Sciences
  Center
Dallas, Texas

Albert Zlotnik, Ph.D.
Staff Scientist
DNAX Research Institute of Molecular
  Cellular Biology
Palo Alto, California

TABLE OF CONTENTS

Chapter 1

MECHANISMS OF TRANSMEMBRANE SIGNAL TRANSDUCTION DURING B CELL ACTIVATION

K. Mark Coggeshall, John G. Monroe,
John T. Ransom, and John C. Cambier

TABLE OF CONTENTS*

---

*   Abbreviations used: PKC, protein kinase C; PMA, phorbol myristate acetate; DAG, diacylglycerol; PLC, phospholipase C; IP, inositol 1-monophosphate; IP$_2$, inositol 1,4-bisphosphate; IP$_3$, inositol 1,4,5-trisphosphate; PI, phosphatidylinositol; PIP, phosphatidylinositol 4-phosphate; PIP$_2$, phosphatidylinositol 4,5-bis-phosphate; PA, phosphatidic acid; PS, phosphatidylserine; PC, phosphatidylcholine; BCGF$_1$, B cell growth factor 1.

I. INTRODUCTION

It has become clear over the past decade that generation of humoral immune responses to most antigens requires the productive interaction of at least three cell types, a number of soluble cell-derived effector molecules, and antigen. The initial B cell ligand interaction in the generation of thymus-dependent immune responses appears to be antigen binding by membrane immunoglobulin (mIg) of two classes, mIgM and mIgD.[1] The effect of antigen binding is twofold. MIg cross-linking leads to antigen endocytosis, processing, and reexpression.[2] Apparently independent of endocytosis, mIg cross-linking results in generation and transmembrane transduction of signals which drive the quiescent G_0 B cell into a poised state characterized by up to a 50-fold higher expression of surface Ia antigen.[3,4] In this state, which we refer to as G_0*, B cells are optimally prepared to interact with Ia region-restricted and antigen-specific helper T cells.[5] The interaction of G_0* B cells with antigen-specific Ia-restricted T cells results in a further cascade of poorly defined intracellular events which culminates in transition of these cells into the G_1 phase of the cell cycle or blastogenesis.[6] In the presence of nonantigen-specific factors which support proliferation, including IL2, IL1, and BCGF(s), and differentiation, including IFNγ, BCDFμ, and BCDFγ, antibody-secreting cells are generated (for a review, see Chapter 4).

As the preceding discussion illustrates, B cells must have the ability to distinguish and respond differently to a large number of external regulatory species. This presents the cell with a significant dilemma since there are probably a limited number of biochemical mechanisms by which signals generated at the cell surface may be transduced. As discussed below, the ability of the B cell to distinguish among regulatory species is probably determined at multiple levels. These include differential receptor expression, differential receptor coupling to specific second-messenger generating systems, and differential cellular programming for specific responses to specific second messengers. At the most primary level, receptors for regulatory factors may be differentially expressed at specific stages of B cell proliferation or differentiation. For example, growth factor receptors, e.g., IL2 receptors, are not expressed at detectable levels until cells reach G_1.[7] In the case of the lymphokine originally designated $BCGF_1$, receptors are obviously expressed on G_0 B cells since these cells respond to $BCGF_1$ by increasing in surface Ia expression.[8] Although relative isotype distribution changes, cell surface immunoglobulin is apparently expressed throughout proliferation and differentiation until the plasma cell stage is reached.[9]

The utilization of varying biochemical coupling mechanisms to transduce signals which are generated by binding of ligands to different receptors provides a second mechanistic level at which differential responsiveness to regulatory species is generated. For example, antigen-mIg-mediated signals are transduced via activation of polyphosphatidylinositol hydrolysis leading to intracellular free calcium mobilization and activation of a Ca^{++}- and phospholipid-dependent protein kinase, protein kinase C.[10-13] However, signals generated as a result of $BCGF_1$ binding are transduced by an, as yet, undefined mechanism which clearly does not involve activation of PI metabolism, calcium mobilization, or membrane depolarization (see below).

Finally, in some instances, cells in different phenotypic or differentiative states respond differently to the same second messenger, indicating that their phenotype may include a certain inherent programming of responses to specific second messengers. For example, while normal quiescent B cells respond to protein kinase C activators, e.g., phorbol myristate acetate, by increased expression of Ia antigen, examination of a series of B cell lymphomas revealed patterned responses involving up and down regulation of IgM, IgD, and DR expression following stimulation with phorbol diesters.[14] These data suggest that lymphomas "frozen" in specific differentiative states respond differently to activation of the protein kinase C second-messenger system. Differential programming, presumably, also occurs in normal B cells undergoing an immune response. For example, cells in G_0 might respond to generation of the same second messenger differently than lymphoblasts.

It is clear that regulation of lymphocyte activation, proliferation, and differentiation is complex and multifaceted, involving more than simply presence or absence of the appropriate regulatory species. Appropriate receptors must be expressed, receptors must be coupled to appropriate biochemical transduction mechanisms and the cells must be programmed to respond appropriately to the transduced signal. It is the goal of this report to review our current understanding of the transduction mechanisms which may be operative during B cell activation. To this end we will first discuss mechanisms which are potentially operative and then review our recent findings regarding signal transduction via mIg and receptors for $BCGF_1$.

II. TRANSMEMBRANE SIGNALING MECHANISMS

Antigens, mitogens, and cytokines which regulate B lymphocyte physiology all appear to exercise their effects by interacting with cell surface receptor or acceptor molecules rather than by crossing the plasma membrane. The information or "signal" generated by the interaction of ligand with receptor must be rapidly communicated to the interior of the cell so the appropriate response can be made. This involves biochemical transduction of the signal generating intracellular "second messengers". The concept of a second messenger was originally proposed by Sutherland and Rall based upon their studies of the effects of adrenaline on liver[15] and has been, subsequently, applied to a wide variety of cell activation systems. The notion of a second messenger is especially attractive when applied to systems in which ligand receptor interactions lead to rapid alterations in cell physiology. Rapid alterations in physiology such as membrane depolarization by anti-Ig-stimulated B cells can occur only by modification of existing enzymes and proteins, since they occur before detectable increases in RNA and protein synthesis. Covalent modifications which might radically change the biologic activities of substrates, such as phosphorylation, methylation, or adenylation, are the most likely to be operative in these systems. Similarly, noncovalent modifications such as proteolytic cleavage or protein-protein interactions which alter biologic activity may also be important intermediary steps in signal transduction in some instances.

A. Protein Phosphorylation

Recently, much interest has been focused on the role of protein phosphorylation in transmembrane signaling. A great diversity of cellular proteins may be influenced by cycles of phosphorylation and dephosphorylation at serine, threonine, and tyrosine residues. Most cellular phosphoproteins are serine phosphorylated, while succeedingly less phosphorylation occurs at threonine and tyrosine residues, respectively. In fibroblasts about 90% of phosphoprotein is phosphoserine, 10% is phosphothreonine, and 0.002% is phosphotyrosine.[16] Stimulation of cells with epidermal growth factor causes a rapid tenfold increase in tyrosine phosphorylation.[17] Similar rapid increases in tyrosine phosphorylation are induced in a variety of other tissues by other growth factors, including platelet-derived growth factor, insulin, and somatomedin C.[18] Cell surface receptors for these factors have, in some cases, been shown to be protein kinases,[19,20] as have certain oncogenes including PP60 SRC.[18] Other independent protein kinases have been defined and are, in some instances, implicated in specific transduction systems. These kinases are themselves dependent on specific second messengers for their activity. They include cyclic AMP (cAMP)-dependent protein kinase (protein kinase A), cyclic GMP (cGMP)-dependent protein kinase (protein kinase G), calmodulin-dependent protein kinase, protease-activated protein kinase II (PAK II), and calcium- and phospholipid-dependent protein kinase (protein kinase C). These kinases are not tyrosine specific, but phosphorylate primarily, perhaps exclusively, at serine and threonine residues.

B. Cyclic Nucleotides

A number of studies have been directed toward the role of calcium, cAMP, and cGMP, presumably acting via protein kinases in lymphocyte activation. These studies, which have been undertaken using mitogen-stimulated lymphocytes, have resulted in opposing conclusions. Parker[21] has concluded that the mitogenic response of human peripheral blood lymphocytes (HPBL) to Con A and PHA is due to a rise in intracellular cAMP, while studies by Coffey et al.[22] suggest that mitogenesis is due to a rise in cGMP and that cAMP actually inhibits mitogenic responses.

Parker et al.[21] have attempted to resolve the apparent conflicting observations that some agents which elevate intracellular cAMP inhibit lymphocyte activation, while other cAMP elevating agents enhance activation. These investigators have suggested that multiple forms of adenylate cyclase exist in various cellular organelles. In support of this, they have shown that treatment of HPBL with the inhibitory reagents PGE or isoproterenol[23,24] raises cytoplasmic and nuclear cAMP levels, respectively. However, the mitogens Con A and PHA apparently activate an adenylate cyclase confined to the plasma membrane.[23,24] Subsequently, it was shown that mitogens and exogenous cAMP induce similar protein phosphorylation patterns which are distinct from phosphorylation patterns obtained from PGE_1 or isoproterenol-treated HPBL.[21,25,26] In view of the apparent correlation between mitogenesis, rises in membrane-associated cAMP, and protein phosphorylation patterns, Parker has concluded[21] that lymphocyte activation is regulated by a cAMP-dependent protein kinase system.

In opposition to this conclusion, results of Coffey et al.[22] have led to the notion that a rise in cGMP is responsible for mitogenesis in HPBL. These investigators have failed to observe increases in cAMP in HPBL using the same mitogens as in the studies of Parker et al. described above. They note, however, that Con A and the calcium ionophore A23187 induce modest rises (two- to threefold) of cAMP in HPBL, but only at very high concentrations. In contrast, Atkinson et al.[27] have been unable to demonstrate consistent increases in cGMP in mitogen-treated HPBL. Coffey et al.[28] have attributed the inability to demonstrate increases in cGMP to the lack of appropriate sample preparation prior to assay for cyclic nucleotide. However, Atkinson et al.[27] have observed no mitogen-induced increases in cGMP using several purification schemes, including the one suggested by Coffey et al.[22,28] Although this dilemma has not been resolved, it is generally agreed that elevation of intracellular cAMP levels early during mitogen stimulation is inhibitory for activation.

Aside from this dilemma, there is strong evidence that cAMP functions late following stimulation to promote entry of mitogen-activated cells into S phase of the cell cycle. Thus, Foker et al.[29] have demonstrated that Con A-stimulated cells increase approximately threefold in cAMP just prior to DNA synthesis. The increase is transient and returns to basal levels before thymidine incorporation is maximal. In support of the notion that this event regulates entry into S, Foker and co-workers showed that prolonging the rise in cAMP by the addition of exogenous cAMP delays entry into S, but once the exogenous source is removed, the cells incorporate thymidine.[29] Similarly, preventing the increase in cAMP by fusion of lymphocytes with red cell ghosts containing monoclonal anticAMP antibodies[30] or addition of indomethacin blocks the entry into S phase and, in the case of indomethacin, the effect is restored once the inhibitor is removed.[29] These studies are similar to results obtained in fibroblasts which also reveal a surge of cAMP prior to entry into S phase,[31] but contrast with findings of

Katz et al.,[32] who detected no G_1 phase rise in cAMP in activated lymph node cells. In spite of this, the general consensus at present is that cAMP is probably important late in mitogenesis prior to entry into S.

In support of a role for cGMP in lymphocyte activation, Coffey et al.[22] have called attention to several nuclear events that occur in mitogenesis which are regulated by cGMP. These include cGMP-induced phosphorylation of nuclear proteins similar to that induced by Con A,[33] activation of RNA polymerase and RNA synthesis by cGMP,[34] and activation of enzymes by cGMP[35] that are similarly activated in mitogen-induced lymphocyte stimulation. This notion is further supported by studies using agents that increase intracellular cGMP such as acetylcholine, imidazole, and exogenous cGMP. Such agents are not themselves mitogenic but synergize with mitogens to enhance thymidine incorporation.[32,36,37]

The fact that cGMP affects events such as RNA and DNA synthesis that are distal in the activation process suggests that this mediator may be involved in later events as the evidence suggests for cAMP. Studies of Katz et al.[32,38] show that macrophage-depleted, mitogen-stimulated lymph node cells become blocked late in G_1, but are driven into S after the addition of interleukin 1 (IL1). The effect of IL1 could be replaced by exogenous cGMP or by pharmacological agents that induce increased intracellular cGMP. Finally, Katz et al. showed that IL1 induced an increase in cGMP in mitogen-stimulated cells, and this increase was shown to occur at approximately the same time as the IL1 requirement[32,38] is seen. These investigators concluded that mitogen induces a transition from G_0 to G_1 in mitogen-sensitive cells and that IL1 and, by extension, cGMP promote entry into S.

An additional observation which suggests that cGMP is not involved in early activation signals comes from studies of Goldberg et al.[39] These workers investigated the cellular distribution of guanylate cyclase, the enzyme responsible for the synthesis of cGMP, and found that it is not a plasma membrane-bound enzyme. Much of the activity was shown to be confined to the nucleus. This result is consistent with the observations of Coffey et al.[22] who showed that many of the cGMP-regulated events occur in the nucleus. These events appear to involve macromolecular metabolism such as RNA, DNA, and protein synthesis. Since macromolecular synthetic events appear relatively late in mitogen-stimulated lymphocytes, these observations further support the notion that cGMP via protein kinase G regulates later events.

C. Calcium

A number of studies have shown that lymphocyte activation by mitogens is accompanied by a rapid rise in intracellular Ca^{++} and has led to speculation that the mitogenic signal is transduced via a calcium-regulated kinase.[40] In addition to these observations, Parker[41] has shown a positive correlation between the mitogenic potential of various agents and their ability to induce increases in intracellular Ca^{++}.

Studies using Ca^{++} chelating reagents have shown that extracellular Ca^{++} is required for late events such as thymidine incorporation,[42-44] as well as earlier events such as hexose[45] and amino acid uptake[42,46] and RNA synthesis.[47] These results support the hypothesis that Ca^{++} may play a role in transmembrane signaling. In contrast, however, several studies[43,44] have shown that extracellular Ca^{++} is required only for transition into S phase and that progression through the cell cycle up to this point can be made in the presence of Ca^{++} chelators. However, these experiments employed relatively low concentrations of Ca^{++} chelators. Other experiments[45,47] have shown that Ca^{++} concentrations as low as 50 μM are sufficient to support early events. Our own unpublished data show that sufficient Ca^{++} would be present in the media to support early events in the presence of the concentrations of chelators employed. In addition, we have observed that most Ca^{++} needed for signaling is derived from intracellular Ca^{++} stores.[13]

Artificial induction of a rise in intracellular Ca^{++} by the ionophore A23187 has been shown to be an efficient mitogen in some systems, but the response varies greatly with species. In porcine and human lymphocytes, A23187 is less effective than PHA or Con A with respect to thymidine incorporation,[48-50] but is almost as effective as the mitogens in the activation of RNA and protein synthesis.[50] In rat thymocytes, A23187 is a poor mitogen but a potent stimulator of glucose transport.[51] Resch et al.[52] have provided evidence that A23187 is mitogenic for rabbit B lymphocytes. Toyoshima et al.[53] have shown that while A23187 is a poor mitogen for mouse splenocytes, it will synergize with a late pulse of Con A to induce thymidine incorporation in these cells.

Blitstein-Willinger and Diamantstein[54] have investigated the effects of the calcium channel blocker verapamil on Con A and PHA induced thymidine incorporation. Their results indicated that verapamil inhibits a late step, between G_1 and S phases of the cell cycle. The inhibition did not appear to affect early signals, since cells pulsed with Con A in the presence of verapamil for 24 hr, washed, and recultured with Con A for an additional 24 hr incorporated ^3H-thymidine at 70% of the control (Con A alone for 48 hr). Unfortunately, this study did not show that verapamil blocks early Ca^{++} uptake. The concentration of verapamil used was determined by inhibition of thymidine incorporation, and it is possible that higher verapamil concentrations are required to block early events. Furthermore, verapamil appears only to block voltage-gated Ca^{++} channels.[55] Our recent studies indicate that Ca^{++} mobilization which occurs early in activation occurs by a nonvoltage-dependent mechanism.[13] Thus, the possibility remains that Ca^{++} is involved in the transmembrane signal transduction during lymphocyte mitogenesis.

A difficulty in drawing a general conclusion to the role of Ca^{++} in lymphocyte activation is that the activity of enzymes that generate second messengers themselves exhibit Ca^{++} dependency. Thus, Coffey et al.[22,28] have shown that the effects of cGMP in the lymphocyte nucleus require the presence of calcium. This requirement may be due to the ability of calmodulin to activate cGMP-dependent protein kinase[56,57] or may reflect the ability of Ca^{++} to activate guanyl cyclase.[22] Calmodulin has been shown to be capable of activating several species of protein kinases, although the substrate or phosphate acceptor protein for these kinases appears to be relatively specific.[58] It is difficult, though not impossible, to envision a mechanism by which a protein kinase with a single specific substrate is able to mediate the diverse biochemical and cell biologic events that accompany lymphocyte activation.

In virtually all systems, including mitogen stimulation of lymphocytes which appear to employ Ca^{++} as an intracellular second messenger, ligand binding also evokes a rapid turnover of phosphatidylinositol, a minor constituent (~5 to 7%) of membrane phospholipids. As will be discussed in much greater detail below, hydrolysis of phosphoinositides liberates diacylglycerol and polyphosphoinositol. Recent evidence suggests that polyphosphoinositol acts in promoting mobilization of Ca^{++} from intracellular stores[59] and that this Ca^{++} acts in concert with diacylglycerol to activate protein kinase C leading to the biologic response.

III. MEMBRANE IMMUNOGLOBULIN-MEDIATED TRANSMEMBRANE SIGNALING

During the past 4 years we have conducted extensive studies of signal transduction by B cell immunoglobulin following antigen or antiimmunoglobulin antibody binding. These studies were conducted under conditions of or mimicking thymus-dependent antigen stimulation in which neither blastogenesis nor proliferation are induced. Initially, these studies were hampered by a lack of a readily assayable and relevant biological marker of signal transduction. However, in 1981, Mond et al.[3] demonstrated that cross-linking of mIg in vitro or in vivo leads to a five- to tenfold increase in expression of cell surface Ia antigens first detectable within 6 to 8 hr following stimulation. Results from several laboratories suggest that this change in mIa expression may be a, or perhaps the, most important cellular change during G_0 to G_0^* transition. Thus, Ia antigen expression by B cells is of most obvious importance, because these molecules act as recognition/restriction elements for effective collaboration between T and B cells in generation of antibody responses (for review see Singer and Hodes[60]). Data of Henry et al.[61] and Bottomly et al.[62] further suggest that B cells which express *high levels* of Ia interact most efficiently with Ia-restricted helper T cells in generation of antibody responses. Further, our recent studies indicate that stimulation of normal B cells to enter the high Ia state greatly increases their ability to present antigens to T cells resulting in IL-2 production.[8] Using a recently developed flow cytometric method to measure T cell-B cell conjugate formation, we have demonstrated a direct correlation between the level of Ia expression by B cells and their ability to bind to T cells in the presence of appropriate antigen (Figure 1). Thus, in systems which involve cognate T cell-B cell interactions, the effect of increased Ia expression appears minimally to promote a more avid and, thus, more productive cellular interaction.

Induction of increased Ia antigen expression by antireceptor is inhibited by both actinomycin D[9] and cycloheximide,[3] suggesting that this process involves new DNA transcription and RNA translation. Increased expression of complement receptors also appears to accompany G_0 to G_0^* transition.[63] In other respects thus far examined, G_0^* B cells appear phenotypically identical to G_0 cells. They are small (4.5 to 5.5 μm diameter), contain a G_0 level of RNA as defined by acridine orange cell cycle analysis, and express G_0 cell levels of cell surface Qa, H2-D, H3K, IgM, and IgD antigens.[103]

A. Membrane Depolarization and mIa Expression

Using changes in mIa antigen expression as an indicator of G_0 to G_0^* transition, we have begun to define causal relationships between cell physiologic changes resultant from mIg cross-linking and transition to G_0^*. Particularly informative in these studies has been a change in cellular uptake of the carbocyanine dye $DiOC_5[3]$ following receptor cross-linking. Decreased uptake of this dye, which is detectable within the first minutes of stimulation,[64] appears indicative of plasma membrane depolarization. In support of this interpretation is the fact that equivalent changes in staining occur following membrane depolarization induced by elevation of extracellular K^+ to 50 mM.[64] Furthermore, antireceptor antibody induction of this response is blocked in the presence of the potassium ionophore valinomycin.[4] Membrane depolarization also accompanies mitogenic stimulation of B cells by LPS and T cells by Con A.[65] In the case of B cells stimulated with LPS, antigen, or antireceptor antibody, membrane depolarization which is detectable within 3 min and maximal within 1 hr is always followed by increased Ia antigen expression, suggesting that these events may be linked. This possibility is further supported by recent findings that high extracellular K^+ (50 mM) induces not only membrane depolarization, but also increased Ia expression.[4] Further, valinomycin blocks induction of both membrane depolarization and increased Ia expression by antireceptor antibodies.[4] These findings suggest that membrane depolarization may be a key intermediary event in mIg-mediated signaling, which occurs before significant branching of the transduction cascade.

B. Protein Kinase C and Membrane Depolarization

Having defined membrane depolarization as an early indicator of mIg-mediated signal transduction and as an event which appears to be causally related to increased Ia expression, the next logical step was to investigate the mechanism of coupling between receptor cross-linking and the alteration in ion flux manifest by membrane depolari-

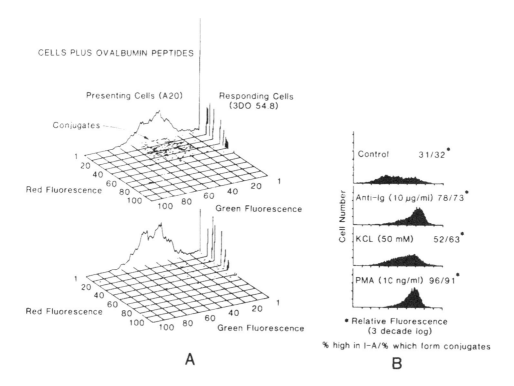

FIGURE 1. B cell-T cell conjugate formation. Rhodamine isothiocyanate-labeled T hybridoma cells (3D054.8) were mixed with fluorescein isothiocyanate-labeled B cells of appropriate I-A haplotype plus tryptic digest of ovalbumin (100 μg/mℓ). Either A20 lymphoma cells (left panel) or normal B cells were used. Cells were mixed at a final cell concentration of 10^6 in 100 μℓ (at a 1:1 ratio in the left panel) and incubated in complete medium for 4 hr at 37° in 5% CO_2 before being suspended in 1 mℓ warm medium and assayed by two-color fluorescence analysis using a Cytofluorograf 50H. Typical results are shown in the isometric display (left panel) in which particles displaying both red and green fluorescence represent conjugates. Conjugates form only when appropriate antigen (Ova peptide) is present in the system. In the right panel are displayed the I-A expression profiles (immunofluorescence) of normal B cells after stimulation in vitro for 24 hr with antiimmunoglobulin, KCl, or PMA, as well as that of cultured, unstimulated cells. These populations were analyzed by the conjugate assay using a 5:1 ratio of T cells (3D054.8) to B cells. At right are shown the percent of B cells expressing high I-A (numerator) and the percent which form conjugates (denominator).

zation. Studies of protein kinase C (PKC) and its activation by phorbol diester analogues provided, perhaps, the most important insight into the operative mechanism. Ogawa et al.[66] described this calcium-regulated enzyme which occurs in high concentration in lymphoid tissues. A variety of subsequent reports suggested that the biologic activities of phorbol esters, including promotion of tumorigenesis, induction of platelet aggregation, promyelocyte differentiation, and lymphocyte activation, may be mediated via this protein kinase. The enzyme appears to be the cellular "receptor" for phorbol and is directly activated by biologically active phorbol diesters.[10,67,68] Lindsten et al.[63] and Monroe et al.[10] have documented phorbol diester induction of increased Ia expression by normal B cells. We[10] have further shown that a strong correlation exists between the ability of phorbol diester analogues to activate mouse B lymphocyte-derived PKC in cell-free systems and to stimulate membrane depolarization and increased Ia antigen expression by intact B cells. These observations led us to hypothesize that PKC activation may be a coupling event between receptor immunoglobulin cross-linking and membrane depolarization. Consistent with this hypothesis are observations that new protein phosphorylation is detectable early during lymphocyte mitogenesis[69,70] and that K+ flux may be regulated by phosphorylation in other tissues.[71,73] A role of

PKC in receptor immunoglobulin-mediated B cell signaling is further supported by recent observations that patterns of new protein phosphorylation detectable within 15 min following stimulation of B cells with antireceptor antibody appear similar, if not identical, to patterns observed following stimulation with phorbol myristate acetate (Figure 2). These patterns demonstrate the appearance of a new phosphoprotein of approximately 20,000 daltons following PMA or antiimmunoglobulin stimulation of B cells.

C. Phosphatidylinositol Metabolism and mIg Signaling

In view of data implicating a role for PKC in mIg-mediated signaling resulting in G_0 to G_0^* transition, we have begun to study the mechanism by which surface Ig cross-linking and PKC activation may be coupled. Kishimoto et al.[74] have proposed that the physiologic activator of PKC is diacylglycerol (DAG). This hypothesis is based upon findings that DAG, a product of membrane phosphoinositide hydrolysis which accompanies receptor binding in a number of systems (for review see Mishell and Kirk[75]), is, like phorbol diester, a potent activator of isolated PKC. Recently, both exogenous DAG and phorbol diesters have been shown to induce platelet aggregation[76] and phosphorylation of a 52,000-kd protein in chick embryo fibroblasts.[77] We have shown that exogenous DAG either in the form of 1-oleyl-2-acetylglycerol or diolein induce both membrane depolarization and increased Ia expression by normal B cells.[11] If increased diacylglycerol generation from phosphoinositides is an important intermediary event in mIg-mediated signal transduction, increased PI metabolism should be detectable early following antireceptor stimulation of B cells. It should be noted that it is possible that the diacylglycerol important in activation could be generated from some source other than phosphoinositides. It is most likely, however, that second-messenger DAG arises via the "PI cycle" illustrated below (Figure 3) by hydrolysis of phosphoinositides by phospholipase C (PLC). Early studies by Michell and others[75] documented increased phosphoinositide metabolism following lymphocyte stimulation with conventional mitogens, and Maino et al.[78] observed increased PI synthesis following stimulation of porcine B cells with anti-Ig antibodies. We have recently documented a 20 to 30% loss of PI and a fourfold increase in phosphatidic acid generation detectable within 1 to 2 min of stimulation by antireceptor antibody, which is followed within 10 min by increased PI resynthesis.[12] Preliminary studies indicate that concomitant with the loss in PI, an increase in release of inositol 1-monophosphate (IP), inositol 1,4-bisphosphate (IP_2), and inositol 1,4,5-trisphosphate (IP_3) is detectable.

These findings are consistent with mIg-mediated signal transduction occurring by sequential initiation of phosphoinositide hydrolysis liberating DAG, DAG activation of protein kinase C, and a phosphorylation-mediated alteration of ion distribution. As discussed below, IP_3 released during this process may also play an important role in protein kinase C activation via its effect on intracellular free Ca^{++} levels.

D. Ca^{++} Mobilization and B Cell Activation

Recent studies by Pozzan et al.[79] have demonstrated a rapid increase in intracellular free calcium levels following stimulation of B lymphocytes by antireceptor antibody. We have observed that this increase in free calcium occurs at all ligand concentrations which induce membrane depolarization and increased Ia expression.[13] Further, in B cells, the calcium ionophore A23187 is a potent inducer of membrane depolarization and increased Ia expression in the absence of significant PI hydrolysis.[13] Thus, these data suggest that the increase in intracellular free calcium which occurs during receptor Ig-mediated signal transduction could itself drive membrane depolarization and transition to G_0^*. Two obvious questions arise. First, what is the mechanism by which receptor cross-linking leads to increased intracellular free Ca^{++}? Second, how do intra-

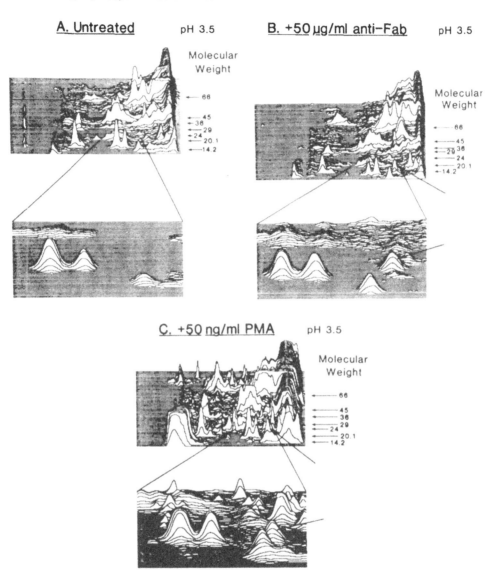

FIGURE 2.   Two-dimensional gel electrophoretic analysis of B lymphocyte phosphoprotein. B cells prela-beled with ^{32}Pi as described[13] were incubated 15 min with no addition (A), 50 μg/mℓ rabbit antimouse Fab (B), or 50 nM phorbol myristate acetate (C). Phosphoproteins were extracted and analyzed as described[102] and the resolved phosphoproteins were detected by autoradiography. The film was then scanned by a Zeinch scanning densitometer equipped with an Apple IIE® desk-top computer for data analysis. The section of the gel showing a qualitative difference is outlined and shown on an expanded scale beneath. Both PMA and anti-Fab-treated cells reveal a new phosphoprotein (marked by arrows) in this area of the gel. This protein has a pI between 4.5 and 5.0 and a molecular weight near 20,000 daltons.

cellular free Ca^{++} levels modulate membrane potential and, if it is not related to PKC activity, why does the cell apparently utilize redundant mechanisms to drive cellular changes manifest by membrane depolarization? Berridge has suggested that the mobi-lization of intracellular calcium is mediated by inositol 1,4,5-triphosphate (IP$_3$) re-leased by hydrolysis of phosphatidylinositol 4,5-bisphosphate.[80] This hypothesis is sup-ported by studies demonstrating specific hydrolysis of phosphatidylinositol 4,5-bisphosphate (PIP$_2$) and phosphatidylinositol 4-phosphate (PIP) very early during sig-naling in platelets,[81] parotid tissue,[82] hepatocytes,[83,84] and pancreatic acinar cells.[85]

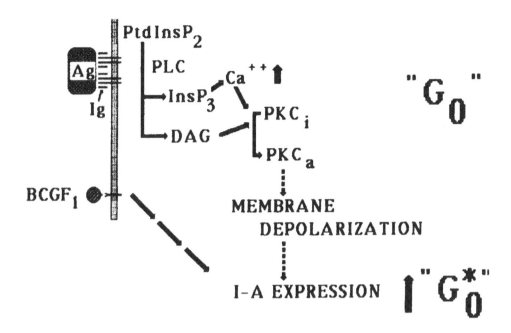

FIGURE 3. The phosphatidylinositol cycle. Ag, Antigen; Ig, surface immunoglobulin; PtdInsP$_2$, phosphatidylinositol 4,5-bisphosphate; PLC, phospholipase C; InsP$_3$, inositol 1,4,5-trisphosphate; DAG, diacylglycerol; PKCi and PKCa, protein kinase C inactive and active; BCGF$_1$, B cell growth factor 1.

Rapid release of inositol 1,4-bisphosphate (IP$_2$) and inositol 1, 4, 5.-trisphosphate (IP$_3$) has been demonstrated to accompany PIP and PIP$_2$ breakdown following appropriate stimulation in insect salivary glands, rat brain, parotid tissue, hepatocytes, B cells,[103] and platelets.[80,86-88] Further, exogenous IP$_3$ has recently been shown to trigger mobilization of intracellular calcium in pancreatic acinar cells, hepatocytes, and rat insulinoma.[59,88,89] Thus, results are consistent with the possibility that hydrolysis of PIP$_2$ following receptor binding could mediate mobilization of intracellular free Ca^{++} in many systems, including B lymphocytes stimulated with their specific antigen.

The answer to the second question, i.e., how does the calcium mobilization response modulate membrane potential, is less apparent. Hoffmann and Majerus[90] have shown that Ca^{++} can modulate PI hydrolysis, amplifying PI hydrolysis by phospholipase C (PLC). Thus, Ca^{++} mobilization could lead to amplification of the PI hydrolytic response resulting in potentiation of PKC activity. Similarly, since protein kinase C is a calcium-regulated enzyme,[66] Ca^{++} could directly amplify its activity. Our studies in lymphocytes have demonstrated a clear synergy between calcium ionophore A23187 and phorbol myristate acetate in stimulating membrane depolarization and increased Ia expression, suggesting that during a physiologic response diglyceride and Ca^{++} may act in concert to activate protein kinase C.[13] Synergy of these agents has also been demonstrated in activation of serotonin release by platelets.[91]

E. The Model

Based upon the data discussed above it is possible to develop a working model for transmembrane signaling by mIg.[8] As shown above (Figure 3), we hypothesize that cross-linking of mIg initiates hydrolysis of PI, but perhaps more importantly, hydrolysis of PIP$_2$ resulting in the generation of DAG and IP$_3$, which in turn liberates Ca^{++} from intracellular stores. DAG and free Ca^{++} act in concert to activate protein kinase C. PKC-mediated phosphorylation leads to an alteration in ion flux manifest by membrane depolarization. This altered ion flux, through unknown intermediary steps, appears to mediate an increase in Ia gene transcription, translation, and membrane expression. Entry into cycle does not follow unless an appropriate second signal is provided by mitogenic moieties or, in the case of thymus-dependent antigens, an antigen-specific and Ia-recognizing T cell or its soluble product.

In support of the validity of this cascade, following cross-linking of receptor immunoglobulin is the documentation of PI hydrolysis,[12] phosphatidic acid generation,[12] elevation of intracellular free calcium,[13] new protein phosphorylation (Figure 2) membrane depolarization,[64] and increased Ia expression[64] occurring in a temporal sequence consistent with the above model following stimulation of B cells by F(ab')$_2$ anti-Ig antibodies (Figure 4). In support of the ability of this cascade to drive increased Ia transcription, translation and expression are observations that exogenous phospholipase C,[11] diacylglycerol,[11] A23187 plus Ca^{++},[13] phorbol diester,[10] and 50 mM K^{+4} each stimulate membrane depolarization and increased Ia antigen expression (Table 1 and 2). Finally, as shown in Table 2, the fact that cAMP plus theophylline, used under conditions which block PI metabolism in this system,[12] also prevented anti-Ig- (but not PMA-) induced membrane depolarization and increased Ia expression[11] suggests that this cascade is necessary for receptor Ig cross-linking-induced increased IA expression.

The role of calcium in these events is, as yet, unclear. As mentioned above, it appears that increased intracellular calcium and diacylglycerol, both generated as a consequence of receptor occupancy, synergize in the activation of protein kinase C. However, in their studies of protein kinase C, Nishizuka[92] has observed that DAG serves to lower the Ca^{++} concentration required by protein kinase C, thus leading to the hypothesis that sufficient Ca^{++} levels exist in the resting cell to induce PKC activity when in the presence of DAG generated from PI. We have observed that depolarization occurs in response to anti-Ig antibodies in nominally Ca^{++}-free media or in the presence of the Ca^{++} chelater EGTA. However, such experiments are difficult to interpret, since under these conditions a rise in intracellular Ca^{++} still occurs, reflecting release of Ca^{++} from intracellular stores. Because of the adverse effects of Ca^{++}-free medium on lymphocyte viability, we have been unable to define the extracellular calcium requirement for increased Ia expression. Experiments with the calcium channel blocker D600 (Figure 5) have revealed that concentration up to 100 μM fail to block anti-Fab-induced hyper-I-A antigen expression. However, concentrations of D600 as low as 10 μM appear to inhibit cell cycle transition as measured by thymidine incorporation (Table 3). These data would suggest a role for extracellular Ca^{++} at a point after transition of the B cell to G$_0$*. On the other hand, the observations that the Ca^{++} ionophore A23187 is able to induce both depolarization[13] and increased I-A expression (Figure 6) suggest that Ca^{++} may play a role in these responses. The most logical explanation for these findings is that Ca^{++} mobilization is required at two levels in lymphocyte mitogenesis. That required as a second messenger in mIg-mediated transmembrane signaling arises from intracellular stores by a D600-insensitive mechanism, while that required later for entry into cycle arises via a D600-sensitive mechanism.

Based on their studies of Ca^{++} ionophore-induced platelet activation, Billah and Lapetina[93,94] have attributed aggregation induced by A23187 to Ca^{++}-mediated activation of platelet phospholipase A$_2$. They have suggested that arachidonic acid products generated by the action of phospholipase A$_2$ feedback on the PI cycle, resulting in the generation of PA, presumably, through activation of phospholipase C and DAG generation.[95] If a similar pathway exists in lymphocytes, it would be expected that ionophore-induced depolarization should be sensitive to inhibitors of phospholipase A$_2$ and/or arachidonic acid metabolism. However, the data shown in Table 4 reveal that A23187-induced depolarization is not sensitive to the phospholipase A$_2$ inhibitors quinacrine or bromophenacylbromide, nor to the arachidonic acid metabolism inhibitor indomethacin. These results suggest that the action of A23187 on B cells is not depend-

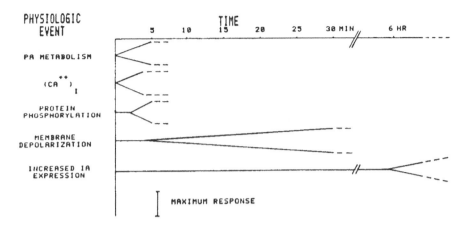

FIGURE 4. Approximate temporal relationships of defined cell biologic events which follow membrane immunoglobulin cross-linking. Events assayed include activation of phosphatydic acid (PA) metabolism, mobilization of intracellular free calcium [(Ca^{++})$_i$], new protein phosphorylation, membrane depolarization, and increased cell surface I-A expression. Divergence of bars indicates occurrence of a detectable response which is maximal at the bar width indicated.

Table 1

MIMICRY OF mIg-MEDIATED SIGNALING BY PHARMACOLOGIC
AGENTS

Inducers	PI hydrolysis	(Ca^{++})$_i$	Protein phosphorylation	Membrane depolarization	Increased IA expression
Anti-Ig	+	+	+	+	+
Phospholipase C (0.5 Ug/mℓ)	$^\triangle$+	+	ND	+	+
Ca^{++} ionophore A23187 (25 nm)	−	$^\triangle$+	ND	+	+
Diacylglycerol (10 μm)	−	−	$^\triangle$ND	+	+
PMA (1 nm)	−	−	$^\triangle$+	+	+
K$^+$ (50 mm)	−	−	ND	$^\triangle$+	+

Note: $^\triangle$ = Theoretical site of action.

Table 2

INHIBITION OF ANTI-Ig-INDUCED TRANSMEMBRANOUS SIGNALING

Inhibitors	PI hydrolysis	(Ca^{++})$_i$	Protein phosphorylation	Membrane depolarization	Increased IA expression
	0	+	+	+	+
dbcAMP (100 μM) + theophylline (1 mm)	$^\triangle$−*	−*	ND	−	−
Valinomycin (10^{-8} M)	+	+	ND	$^\triangle$−	−
Sodium azide (0.2%)	ND	ND	ND	+	ND

Note: $^\triangle$ = Site of action; * = 50—90% inhibition.

ent on phospholipase A$_2$ and subsequent arachidonic acid metabolism. In addition, as shown in Table 5, A23187 does not appear to activate the PI cycle. Thus, data indicate that A23187 does not mediate its effect by activation of PI metabolism, but rather by mediating a direct Ca^{++} signal.

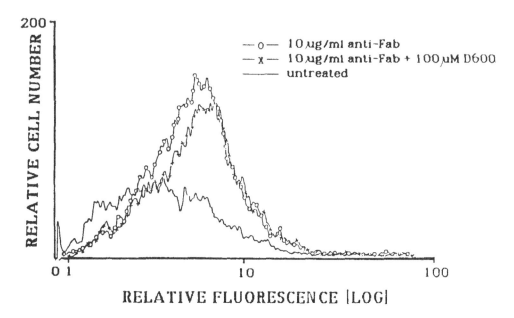

RELATIVE FLUORESCENCE |LOG|

FIGURE 5.   D600 does not block anti-Fab-induced hyper I-A expression. Cells were treated with anti-Fab in the presence or absence of D600, cultured, stained, and analyzed as previously described.[11] Shown are fluorescence histograms of cells treated with 10 μg/ml anti-Fab (—O—), 10 μg/ml anti-Fab in the presence of 100 μM D600 (—X—), and untreated cells (−).

Table 3
D600 INHIBITS ANTI-Fab-INDUCED THYMIDINE
INCORPORATION

Culture conditions[a]	% Maximal thymidine[b] incorporation
Untreated	3
25 μg/ml RaM anti-Fab	100
25 μg/ml RaM anti-Fab + 50 μM D600	60
25 μg/ml RaM anti-Fab + 100 μM D600	8

[a]   2 × 10 B cells per microwell in RPMI + 10% fetal calf serum + 50 μM
2-ME + glutamine, penicillin, and streptomycin. Cells were incubated
for 48 hr at 37°C, 10% CO_2, pulsed with 1 μCi ^{3}H thymidine, and
harvested 16 hr later.
[b]   Normalized to response obtained with 25 μg/ml RaM anti-Fab.

IV. FUTURE DIRECTIONS

Clearly, much remains to be done to define the mIg-mediated transmembrane signaling mechanism. For example, the relative contribution of PI, PIP, and PIP$_2$ hydrolysis in this signaling process and the potential role of inositol phosphates in mediating calcium mobilization have not been defined. The mechanism by which protein kinase C activation leads to alterations in ion transport manifest by membrane depolarization is currently totally obscure. Equally obscure is the mechanism by which altered ion transport manifest by membrane depolarization may lead to new gene expression. Finally, in no system is the molecular basis of coupling of receptor occupancy to PIP$_2$ hydrolysis known. As discussed below, the mIg-B lymphocyte system appears to be an ideal one in which to study this question.

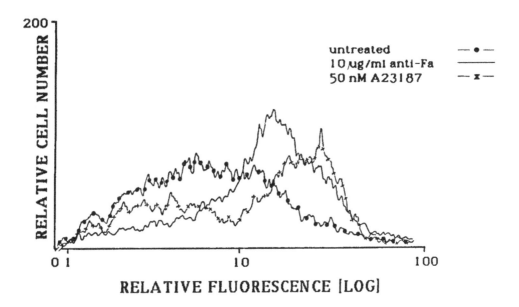

FIGURE 6. A23187 induces hyper I-A expression in B cells. Cells were cultured, stained, and analyzed as previously described.[11] Shown are fluorescence histograms of cells treated with 10 μg/ml anti-Fab (−), 50 nM A23187 (—X—), and untreated cells (—●—).

Table 4

DRUG SENSITIVITY OF A23187-INDUCED
DEPOLARIZATION

Treatment[a]	Relative % of cells depolarized at 2 hr[b]
None	26
300 nM A23187	100
300 nM A23187 + 100 μM quinacrine	94
300 nM A23187 + 10 μM bromophenacyl bromide	100
300 nM A23187 + 50 μM indomethacin	98

[a] Highest drug concentrations tested.
[b] Normalized to % cells depolarized by 300 nM A23187, usually 75—80% of the total B cell population.

Table 5

A23187 DOES NOT INDUCE
PHOSPHATIDYLINOSITOL METABOLISM

Treatment	^{32}P cpm in fraction (fold increase)		
	PA	PI	PC
Untreated	9,000 (1.0)	80,024 (1.0)	49,504 (1.0)
R anti-M Fab	21,704 (2.4)	138,515 (1.7)	54,414 (1.1)
A23187	8,985 (1.0)	83,924 (1.0)	56,034 (1.1)

A. Ig Cross-Linking — PI Hydrolysis Coupling

Initiation of the PIP_2 hydrolysis response and its sequelae obviously depends upon cellular expression of appropriate receptors and coupling of those receptors to hydrolysis of PIP_2. The mechanism of this coupling is a key question now confronting receptor biologists. Transmembrane signaling via membrane immunoglobulin on B cells represents an ideal model to study this question for a number of reasons. Receptor immunoglobulin has been virtually completely defined structurally, including amino acid sequence, membrane orientation, and expression density. Genes encoding these receptors have been cloned and mutants which vary in specific structural domains are being constructed. Cell lines which have been transfected with these mutant genes should provide powerful tools for exploration of the molecular basis of receptor-PIP_2 hydrolysis coupling.

The two most obvious possibilities for the mechanism of receptor-PIP_2 hydrolysis coupling are (1) activation and/or translocation of phospholipase C following receptor cross-linking leading to increased PIP_2 hydrolysis, and (2) recompartmentalization of phospholipids within the membrane leading to local presentation of an effective substrate for existing PLC and subsequent PIP_2 hydrolysis. The first possibility is currently without support and appears doubtful in view of the inability of Siess and Lapetina[96] to demonstrate detectable increased PLC activity or translocation during stimulation of platelet aggregation. Although no direct evidence supports the second hypothesis, Hoffman and Majerus[90] have demonstrated that ratios of other phospholipids in PI-containing substrates profoundly affect the ability of PLC to hydrolyze the substrate. Particularly, enrichment of phosphoinositides and phosphatidylserine (PS) and exclusion of phosphatidylcholine (PC) leads to increased hydrolysis. If the receptors' cytoplasmic termini exhibit higher affinity for PIP_2 and PS than PC, then aggregation of these c termini during receptor cross-linking might lead to generation of an optimal local substrate configuration for PIP_2 hydrolysis by existing phospholipase C. In the case of membrane immunoglobulin which has a C-terminal tripeptide cytoplasmic tail with lysine-valine-lysine,[97] a selective association with PIP_2 and PS might be predicted based upon charge. We are currently using mutants which vary in the amino acid sequence of the cytoplasmic tail to examine this possibility.

Recent studies which document phosphorylation of PI by pp60 SRC, a tyrosine kinase, raise an additional possibility for coupling of receptor occupancy to the PI hydrolysis response.[98] Specifically, ligand binding may trigger PI and PIP phosphorylation raising the rate of flux of PIP_2 through the PI cycle. This may lead to increased availability of PIP_2, which upon hydrolysis during normal turnover triggers Ca^{++} release from intracellular stores and a subsequent biological response.

B. IgM vs. IgD-Mediated Signaling

Based upon teleological arguments and experimental evidence derived from in vitro immunologic tolerance and immune response models, it is generally believed that IgM and IgD transduce differing signals to the cell.[99] However, it appears clear from our recent studies[100] that in quiescent B cells, both membrane IgM and IgD transduce signals resulting in increased PI hydrolysis, PKC activation, membrane depolarization, and increased Ia expression. Other recent studies have shown that while cross-linking of IgM on small B cells by soluble bivalent anti-IgM antibodies stimulates thymidine uptake, cross-linking of IgD by equivalent IgD-specific reagents does not stimulate thymidine uptake.[101] Further, induction of thymidine uptake by IgM cross-linking reagents requires significantly more ligand than induction of the PI hydrolysis response.[100] These data suggest that IgM, but not IgD, may transduce more than simply a PI hydrolysis activating signal. Exploration of this possibility is currently a major emphasis of our laboratory.

C. Transmembrane Signaling by BCGF₁ Receptors

As discussed earlier, recent findings of Roehm et al.[8] and Noelle et al.[6] indicate that B cell growth factor 1 (BCGF₁) stimulates G_0 B cells to increase in surface Ia expres-

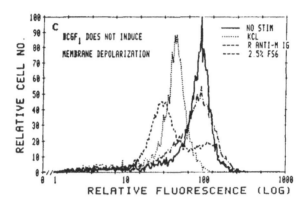

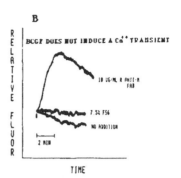

A

BCGF₁ DOES NOT INDUCE PI METABOLISM

³²P CPM IN FRACTION [FOLD INCEASE]

STIMULUS	PA	PI	PC
NONE	7234 [1.0]	77,080 [1.0]	43,414 [1.0]
RAMFab	24,663 [3.4]	153,004 [2.0]	50,233 [1.2]
BCGF₁	7083 [0.98]	79,579 [1.0]	45,363 [1.0]

B

RELATIVE FLUOR

BCGF DOES NOT INDUCE A Ca⁺ TRANSIENT

10 UG/ML R ANTI-R FAB

7.5% FS6

NO ADDITION

2 MIN

TIME

C

RELATIVE CELL NO.

BCGF₁ DOES NOT INDUCE MEMBRANE DEPOLARIZATION

—— NO STIM
········ KCL
- - - R ANTI-M IG
- - - 2.5% FS6

RELATIVE FLUORESCENCE (LOG)

FIGURE 7. BCGF₁ is not coupled to a protein kinase C activation system. (Panel A) Table showing ³²P cpm in various lipid fractions from cells treated with 50 μg/ml anti-Fab, 10% (v/v) FS6 culture supernatant containing BCGF₁, or unstimulated cells. (Panel B) Anti-Fab or FS6 supernatant (BCGF₁) were added to suspensions of B cells loaded with the fluorescent Ca⁺⁺ indicator, Quin 2, and the relative fluorescence measured by standard fluorimetric technique. A rise in Quin 2 fluorescence indicates an increase in intracellular free Ca⁺⁺. The apparently higher level of Ca⁺⁺ in the 7.5% FS6 tracing has been shown to be due only to autofluorescence of the serum protein present in the culture supernatant. (Panel C) B cells (5 × 10⁶/ ml) were cultured for 2 hr in the presence of the reagents indicated. The cells were then diluted to 2.5 × 10⁵/ ml in medium containing the lipophilic, cationic dye DiOC₍₆₎3 (50 nM) and analyzed by flow cytometry after an 8-min equilibration period. Membrane depolarization is indicated by a decrease in relative fluorescence.

sion. This increase is similar in magnitude and time course to that induced by antiimmunoglobulin antibodies. Monoclonal anti-κ and BCGF₁, isolated by ammonium sulfate precipitation and molecular sieving using HPLC, are costimulators of Ia expression when used at suboptimal doses. While the precise role of this factor in generation of immune responses is unclear, these results are consistent with it functioning via an effect on Ia expression to nonspecifically promote the interaction of T cells and antigen-presenting B cells.

We have conducted a series of preliminary studies in an effort to define the biochemical basis of signal transduction by BCGF₁ receptors. Specifically, we have examined the effect of semipurified BCGF₁ (see above) on phosphatidylinositol metabolism, intracellular free calcium levels, and membrane potential. As shown in Figure 7, we have detected no significant change in any of these parameters following stimulation of cells with BCGF₁. We, therefore, conclude that the BCGF₁ receptor is not coupled to the Ia induction response via PI cycle activation, Ca⁺⁺ mobilization, and/or membrane depolarization, i.e., a protein kinase C mechanism. Studies are currently underway in our laboratory to determine whether protein kinase A or protein kinase G, via cyclic nucleotide second messengers, may be involved in the signal transduction mechanism by BCGF₁ receptors.

REFERENCES

1. Goding, J. W., Scott, D. W., and Layton, J. E., Genetics, cellular expression and function of IgD and IgM receptors, *Immunol. Rev.*, 37, 152, 1977.
2. Kakiuchi, T., Chesnut, R. W., and Grey, H. M., B cells as antigen-presenting cells: the requirement for B cell activation, *J. Immunol.*, 131, 109, 1983.
3. Mond, J. J., Seghal, E., Kung, J., and Finkelman, F. D., Increased expression of I-region-associated antigen (Ia) on B cells after crosslinking of surface immunoglobulin, *J. Immunol.*, 127, 881, 1981.
4. Monroe, J. G. and Cambier, J. C., B cell activation. III. B cell plasma membrane depolarization and hyper-Ia antigen expression induced by receptor immunoglobulin cross-linking are coupled, *J. Exp. Med.*, 158, 1589, 1983.
5. Janeway, C. A., Bottomly, K., Babich, J., Conrad, P., Conzen, S., Jones, B., Kaye, J., Katz, M., McVay, L., Murphy, D., and Tite, J., Quantitative variation in Ia antigen expression plays a central role in immune regulation, *Immunol. Today*, 5, 99, 1984.
6. Noelle, R. J., Snow, E. C., Uhr, J. W., and Vitetta, E. S., Activation of antigen-specific B cells: role of T cells, cytokines and antigen in induction of growth and differentiation, *Proc. Natl. Acad. Sci. U.S.A.*, 80, 6628, 1983.
7. Cantrell, D. A. and Smith, K. A., The interleukin-2 T cell system: a new cell growth model, *Science*, 224, 1312, 1984.
8. Roehm, N. W., Leibson, H. J., Zlotnik, A., Kappler, J., Marrack, P., and Cambier, J. C., Interleukin induced increase in Ia expression by normal mouse B cells, *J. Exp. Med.*, 160, 679, 1984.
9. Monroe, J. G. and Cambier, J. C., Cell cycle dependence for expression of membrane associated IgD, IgM and Ia antigen on mitogen-stimulated murine B lymphocytes, *Ann. N.Y. Acad. Sci.*, 399, 238, 1982.
10. Monroe, J. G., Neidel, J. E., and Cambier, J. C., B cell activation. IV. Induction of cell membrane depolarization and hyper-I-A expression by phorbol diesters suggests a role for protein kinase C in murine B lymphocyte activation, *J. Immunol.*, 132, 1472, 1984.
11. Coggeshall, K. M. and Cambier, J. C., B cell activation. VI. Effects of exogenous diglyceride and modulators of phospholipid metabolism suggest a central role for diacylglycerol generation in transmembrane signaling by mIg, *J. Immunol.*, 134, 101, 1985.
12. Coggeshall, K. M. and Cambier, J. C., B cell activation. VIII. Membrane immunoglobulins transduce signals via activation of phosphatidylinositol hydrolysis, *J. Immunol.*, 133, 3382, 1984.
13. Ransom, J. T. and Cambier, J. C., B cell activation. VII. Calcium mobilization is an integral event in mIg mediated transmembrane signaling leading to plasma membrane depolarization and increased I-A expression, *J. Immunol.*, submitted for publication.
14. Roth, P., Halper, J. P., Weinstein, B., and Pernis, B., A phorbol ester tumor promoter induces changes in the expression of immunoglobulins and DR antigens in human lymphoblastoid cells, *J. Immunol.*, 129, 539, 1982.
15. Sutherland, E. W. and Rall, T. W., Fractionation and characterization of a cyclic adenine ribonucleotide formed by tissue particles, *J. Biol. Chem.*, 232, 1077, 1958.
16. Cooper, J. A., Bowen-Pope, D. F., Raines, E., Ross, R., and Hunter, T., Similar effects of platelet derived growth factor and epidermal growth factor on the phosphorylation of tyrosine in cellular proteins, *Cell*, 31, 263, 1982.
17. Gill, G. N. and Lazar, C. S., Increased phosphotyrosine content and inhibition of proliferation in EGF-treated A431 cells, *Nature (London)*, 293, 305, 1981.
18. Hunter, T. and Cooper, J. A., Tyrosine protein kinases and their substrates: an overview, *Adv. Cyclic Nucl. Protein Phosphorylation Res.*, 17, 443, 1984.
19. Roth, R. A. and Cassell, D. J., Insulin receptor: evidence that it is a protein kinase, *Science*, 219, 299, 1983.
20. Bohrow, S. A., Cohen, S., and Staros, J. V., Affinity labeling of the protein kinase associated with the epidermal growth factor receptor in membrane vessicles from A431 cells, *J. Biol. Chem.*, 257, 4019, 1982.
21. Parker, C. W., cAMP and lectin-induced mitogenesis in lymphocytes: possible implications for neoplastic cell growth, *Adv. Cyclic Nucl. Res.*, 9, 647, 1978.
22. Coffey, R. G., Hadden, E. M., Lopez, C., and Hadden, J. W., cGMP and calcium in the initiation of cellular proliferation, *Adv. Cyclic Nucl. Res.*, 9, 661, 1978.
23. Bloom, F. E., Wedner, H. J., and Parker, C. W., The use of antibodies to study cell structure and metabolism, *Pharmacol. Rev.*, 25, 343, 1973.
24. Snider, D. E. and Parker, C. W., Adenylate cyclase activity in lymphocyte subcellular fractions. Characterization of non-nuclear adenylate cyclase, *Biochem. J.*, 162, 473, 1977.
25. Wedner, H. J. and Parker, C. W., Protein phosphorylation in human peripheral lymphocytes. Stimulation by phytomagglutinin and N^6 monobutyryl cyclic AMP, *Biochem. Biophys. Res. Commun.*, 62, 808, 1975.

26. Parker, C. W., Control of lymphocyte function, *N. Engl. J. Med.*, 295, 1180, 1976.
27. Atkinson, J. P., Kelly, J. P., Weiss, A., Wedner, H. J., and Parker, C. W., Enhanced intracellular cGMP concentrations and lectin-induced lymphocyte transformation, *J. Immunol.*, 121, 2282, 1978.
28. Coffey, R. G., Hadden, E. M., and Hadden, J. W., Evidence for cyclic GMP and calcium mediation of lymphocyte activation by mitogens, *J. Immunol.*, 119, 1387, 1977.
29. Foker, J. E., Malkinson, A. M., Sheppard, J. R., and Wang, T., Studies of cyclic AMP metabolism in proliferating lymphocytes, in *The Molecular Basis of Immune Cell Function*, Kaplan, J. G., Ed., Elsevier/North-Holland, Amsterdam, 1979, 57.
30. Ohara, J., Sugi, M., Fujimoto, M., and Watanabe, T., Microinjection of macromolecules into normal murine lymphocytes by means of cell fusion. II. Enhancement and suppression of mitogenic responses by microinjection of monoclonal anti-cyclic AMP into B lymphocytes, *J. Immunol.*, 129, 1227, 1982.
31. MacManus, J. P., Aoynton, A. L., and Whitfield, J. F., Cyclic AMP and calcium as intracycle regulators in control of cell proliferation, *Adv. Cyclic Nucl. Res.*, 9, 485, 1978.
32. Katz, S. P., Kierszenbaum, F., and Waksman, B. H., Mechanism of action of "lymphocyte activating factor" (LAF). III. Evidence that LAF acts on stimulated lymphocytes by raising cyclic GMP in G_1, *J. Immunol.*, 121, 2386, 1978.
33. Johnson, E. M. and Hadden, J. W., Phosphorylation of lymphocyte nuclear acidic proteins: regulation by cyclic nucleotides, *Science*, 187, 1198, 1975.
34. Johnson, L. D. and Hadden, J. W., Cyclic GMP and lymphocyte proliferation: effects on DNA-dependent RNA polymerase I and II activities, *Biochem. Biophys. Res. Commun.*, 66, 1498, 1975.
35. Mizoguchi, Y., Otani, S., Matsui, I., and Morisawa, S., Control of ornithine decarboxylase activity by cyclic nucleotides in phytohemagglutinin induced lymphocyte transformation, *Biochem. Biophys. Res. Commun.*, 66, 328, 1975.
36. Hadden, J. W., Coffey, R. G., Hadden, E. M., Lopez-Corrales, E., and Sunshine, G. H., Effects of levamisole and imidazole on lymphocyte proliferation and cyclic nucleotide levels, *Cell Immunol.*, 20, 98, 1975.
37. Sunshine, G. N., Basch, R. S., Coffey, R. G., Cohen, K. W., Goldstein, G., and Hadden, J. W., Thymopoietin enhances the allogeneic response and cyclic GMP levels of mouse peripheral, thymus-derived lymphocytes, *J. Immunol.*, 120, 1594, 1978.
38. Kierszenbaum, F. and Waksman, B. H., Mechanism of action of "lymphocyte activating factor" (LAF). I. Association of lymphocyte activating factor action with early DNA synthesis in PHA-stimulated lymphocytes, *Immunology*, 33, 663, 1977.
39. Goldberg, N. D., Graff, G., Haddox, M. K., Stephenson, J. H., Glass, D. B., and Moser, M. E., Redox modulation of splenic cell soluble guanylate cyclase activity: activation by hydrophilic and hydrophobic oxidants represented by ascorbic and dehydroascorbic acids, fatty acid hydroperoxides and prostaglandin endoperoxides, *Adv. Cyclic Nucl. Res.*, 9, 101, 1978.
40. Hume, D. A. and Weidmann, M. J., *Mitogenic Lymphocyte Transformation*, Elsevier/North-Holland, New York, 1980, chap. 6.
41. Parker, C. W., Correlation between mitogeneicity and stimulation of Ca^{++} uptake in human lymphocytes, *Biochem. Biophys. Res. Commun.*, 61, 1180, 1974.
42. Greene, W. C., Parker, C. M., and Parker, C. W., Calcium and lymphocyte activation, *Cell. Immunol.*, 25, 74, 1976.
43. Diamantstein, T. and Ulmer, A., The control of immune response *in vitro* by Ca^{++}. II. The Ca^{++}-dependent period during mitogenic stimulation, *Immunology*, 28, 121, 1975.
44. Bard, E., Colwill, R., L'Anglais, R., and Kaplan, J. G., Response of human lymphocytes to mitogen. At what stage is there a Ca^{++} requirement?, *Can. J. Biochem.*, 56, 900, 1978.
45. Yasmeen, D., Laird, A. S., Home, D. A., and Weidemann, M. S., Activation of 3-O-methylglucose transport in rat thymus lymphocytes by concanavalin A. Temperature and calcium ion dependence and sensitivity to puromycin but not to cycloheximide, *Biochim. Biophys. Acta*, 500, 89, 1977.
46. Whitney, R. B. and Sutherland, R. M., Effects of chelating agents on the initial interaction of phytohemagglutinin with lymphocytes and the subsequent stimulation of amino acid uptake, *Biochim. Biophys. Acta*, 298, 790, 1973.
47. Hauser, H., Knippers, R., Schäfer, K. P., Sons, W., and Unsöld, H.-J., Effects of colchicine on RNA synthesis in concanavalin A-stimulated bovine lymphocytes, *Exp. Cell. Res.*, 102, 79, 1976.
48. Wang, J. L., McClain, D. A., and Edelman, G. M., Modulation of lymphocyte mitogenesis, *Proc. Natl. Acad. Sci. U.S.A.*, 72, 1917, 1975.
49. Maino, V. C., Green, N. M., and Crumpton, M. J., The role of calcium ions in initiating transformation of lymphocytes, *Nature (London)*, 251, 324, 1974.
50. Kondorosi, E. and Kay, J. E., The role of calcium in lymphocyte activation by the ionophore A23187 and phytohemagglutinin, *Biochem. Soc. Trans.*, 5, 967, 1977.
51. Reeves, J. P., Calcium-dependent stimulation of 3-O-methylglucose uptake in rat thymocytes by the divalent cation ionophore A23187, *J. Biol. Chem.*, 250, 9428, 1975.

52. Resch, K., Bouillon, D., and Gemsa, D., The activation of lymphocytes by the ionophore A23187, *J. Immunol.*, 120, 1514, 1978.

53. Toyoshima, S., Iwata, M., and Osawa, T., Kinetics of lymphocyte stimulation by concanavalin A, *Nature (London)*, 264, 447, 1976.

54. Blitstein-Willinger, E. and Diamantstein, T., Inhibition by isoptin (a calcium antagonist) of the mitogenic stimulation of lymphocytes prior to the S-phase, *Immunology*, 34, 303, 1978.

55. Tritthart, H., Volkmann, R., Weiss, R., and Fleckenstein, A., Inhibition of calcium dependent action potentials in mammalian myocardium by specific inhibitors of the transmembrane calcium conductivity (Verpamil, D600), in *Recent Advances in Studies on Cardiac Structure and Metabolism*, Vol. 5, Fleckenstein, A. and Dhalla, N. S., Eds., University Park Press, Baltimore, 1975, 27.

56. Kuo, W.-N., Shoji, M., and Kuo, J. F., Stimulatory modulator of guanosine 3':5'-monophosphate-dependent protein kinase from mammalian tissues, *Biochim. Biophys. Acta*, 437, 142, 1976.

57. Kuo, J. F., Shoji, M., and Kuo, W.-N., Mammalian cyclic GMP-dependent protein kinase and its stimulatory modulator, *Adv. Cyclic Nucl. Res.*, 9, 199, 1978.

58. Klee, C. B., Crouch, T. H., and Richman, P. G., Calmodulin, *Ann. Rev. Biochem.*, 49, 489, 1980.

59. Streb, H., Irvine, R. F., Berridge, M. J., and Schultz, I., Release of Ca^{++} from a nonmitochondrial intracellular store in pancreatic acinar cells by inositol-1,4,5-triphosphate, *Nature (London)*, 306, 67, 1983.

60. Singer, A. and Hodes, R. J., Mechanisms of T cell-B cell interaction, in *Annual Reviews of Immunology*, Paul, W. E., Ed., Annual Reviews, Palo Alto, Calif., 1983, 211.

61. Henry, C., Chun, E. L., and Kodlin, D., Expression and function of I region products on immunocompetent cells. II. I region products in T-B interaction, *J. Immunol.*, 119, 774, 1977.

62. Bottomly, K., Jones, B., Kaye, J., and Jones, F., III, Subpopulations of B cells distinguished by cell surface expression of Ia antigens. Correlation of Ia and idiotype during activation by cloned Ia-restricted T cells, *J. Exp. Med.*, 158, 265, 1983.

63. Lindsten, T., Thompson, C. B., Finkelman, F. D., Andersson, B., and Scher, I., Changes in expression of B cell surface markers on complement receptor-positive and complement receptor-negative B cells induced by phorbol myristate acetate, *J. Immunol.*, 132, 235, 1984.

64. Monroe, J. G. and Cambier, J. C., B cell activation. I. Antiimmunoglobulin induced receptor crosslinking results in a decrease in the plasma membrane potential of murine B lymphocytes, *J. Exp. Med.*, 157, 2073, 1983.

65. Kiefer, H., Blume, A. J., and Kaback, H. R., Membrane potential changes during mitogenic stimulation of mouse spleen lymphocytes, *Proc. Natl. Acad. Sci. U.S.A.*, 77, 2200, 1980.

66. Ogawa, Y., Taki, Y., Kawahara, Y., Kimura, S., and Nishizuka, Y., A new possible regulatory system for protein phosphorylation in human peripheral lymphocytes. I. Characterization of a calcium-activated, phospholipid dependent protein kinase, *J. Immunol.*, 127, 1369, 1981.

67. Leach, K. J., James, M. L., and Blumberg, P. M., Characterization of a specific phorbol ester aporeceptor in mouse brain cytosol, *Proc. Natl. Acad. Sci. U.S.A.*, 80, 4208, 1983.

68. Castagna, M., Takai, Y., Kaibuchi, K., Sano, K., Kikkawa, U., and Nishizuka, Y., Direct activation of calcium-activated, phospholipid-dependent protein kinase by tumor promoting phorbol esters, *J. Biol. Chem.*, 257, 7857, 1982.

69. Chaplin, D. D., Wedner, H. J., and Parker, C. W., Protein phosphorylation in human peripheral blood lymphocytes: mitogen-induced increases in protein phosphorylation in intact lymphocytes, *J. Immunol.*, 124, 2390, 1980.

70. Johnstone, A. P., Dubois, J. N., Owen, M. J., and Crumpton, M. J., Phosphoproteins of the lymphocyte plasma membrane, *Biochem. Soc. Trans.*, 8, 182, 1980.

71. O'Brien, T. G. and Krzeminski, K., Phorbol ester inhibits furosemide-sensitive potassium transport in BALB/c 3T3 preadipose cells, *Proc. Natl. Acad. Sci. U.S.A.*, 80, 4334, 1983.

72. Alkon, D. L., Acosta-Urquidi, J., Olds, J., Kugma, G., and Neary, J. T., Protein kinase injection reduces voltage-dependent potassium currents, *Science*, 219, 303, 1983.

73. dePeyer, J. E., Cachelin, A. B., Levitan, I. B., and Reuter, H., Ca^{++}-activated K^+ conductance in internally perfused snail neurons is enhanced by protein phosphorylation, *Proc. Natl. Acad. Sci. U.S.A.*, 79, 4207, 1982.

74. Kishimoto, A., Takai, Y., Mori, T., Kikkawa, U., and Nishizuka, Y., Activation of calcium and phospholipid dependent protein kinase by diacylglycerol, its possible relationship to phosphatidylinositol turnover, *J. Biol. Chem.*, 255, 2273, 1980.

75. Mishell, R. H. and Kirk, C. J., Why is phosphatidylinositol degraded in response to stimulation of certain receptors? *Trends Pharm. Sci.*, 2, 86, 1981.

76. Rink, T. J., Sanchez, A., and Hallan, T. J., Diacylglycerol and phorbol ester stimulate secretion without raising cytoplasmic free calcium in human platelets, *Nature*, 305, 317, 1983.

77. Gilmore, T. and Martin, G. S., Phorbol ester and diacylglycerol induce protein phosphorylation at tyrosine, *Nature*, 306, 487, 1983.

78. Maino, V. C., Hayman, M. J., and Crumpton, M. J., Relationship between enhanced turnover of phosphatidylinositol and lymphocyte activation by mitogens, *Biochem. J.*, 146, 247, 1975.

79. Pozzan, T., Gerslan, P., Tsien, R. Y., and Rink, T. J., Anti-immunoglobulin, cytoplasmic free calcium and capping in B lymphocytes, *J. Cell. Biol.*, 94, 335, 1982.

80. Berridge, M. J., Rapid accumulation of inositol triphosphate reveals that agonists hydrolyze polyphosphoinositides instead of phosphatidylinositol, *Biochem. J.*, 212, 849, 1983.

81. Billah, M. M. and Lapetina, E. G., Rapid decrease of phosphatidylinositol 4,5-bisphosphate in thrombin-stimulated platelets, *J. Biol. Chem.*, 257, 12705, 1982.

82. Downes, C. P. and Wusteman, M. M., Breakdown of polyphosphoinositides and not phosphatidylinositol accounts for muscarinic agonist-stimulated inositol phospholipid metabolism in rat parotid glands, *Biochem. J.*, 216, 633, 1983.

83. Thomas, A. P., Marks, J. S., Coll, K. E., and Williamson, J. R., Quantitative and early kinetics of inositol lipid changes induced by vasopressin in isolated and cultured hepatocytes, *J. Biol. Chem.*, 258, 5716, 1983.

84. Creba, J. A., Downes, C. P., Hawkins, P. T., Brewster, G., Michell, R. H., and Kirk, C. J., Rapid breakdown of phosphatidylinositol 4-phosphate and phosphatidylinositol 4,5-bisphosphate in rat hepatocytes stimulated by vasopressin and other Ca^{++}-mobilizing hormones, *Biochem. J.*, 212, 733, 1983.

85. Weiss, S. J., McKinney, J. S., and Putney, J. W., Jr., Receptor-mediated net breakdown of phosphotidylinositol 4,5-bisphosphate in parotid acinar cells, *Biochem. J.*, 206, 555, 1982.

86. Berridge, M. J., Dawson, R. M. C., Downes, C. P., Heslop, J. P., and Irvine, R. F., Changes in the levels of inositol phosphates after agonist-dependent hydrolysis of membrane phosphoinositides, *Biochem. J.*, 212, 473, 1983.

87. Agranoff, B. W., Murthy, P., and Seguin, E. B., Thrombin-induced phosphodiesteratic cleavage of phosphatidylinositol bisphosphate in human platelets, *J. Biol. Chem.*, 258, 2076, 1983.

88. Thomas, A. P., Alexander, J., and Williamson, J. R., Relationship between inositol phosphate production and the increase of cytosolic free Ca^{++} induced by vasopressin in isolated hepatocytes, *J. Biol. Chem.*, 259, 5574, 1984.

89. Prentki, M., Biden, T. J., Janiic, D., Irvine, R. F., Berridge, M. J., and Wollheim, C. B., Rapid mobilization of Ca^{++} from rat insulinoma microsomes by inositol-1,4,5-trisphosphate, *Nature (London)*, 309, 562, 1984.

90. Hoffman, S. L. and Majerus, P. W., Modulation of phosphatidylinositol-specific phospholipase C activity by phospholipid interactions, diglycerides and calcium ions, *J. Biol. Chem.*, 257, 14359, 1983.

91. Yamanishi, J., Takai, Y., Kaibuchi, K., Sano, K., Castagna, M., and Nishizuka, Y., Synergistic functions of phorbol esters and calcium in serotonin release from human platelets, *Biochem. Biophys. Res. Commun.*, 112, 778, 1983.

92. Nishizuka, Y., The role of protein kinase C in cell surface signal transduction and tumor promotion, *Nature (London)*, 308, 693, 1984.

93. Billah, M. M. and Lapetina, E. G., Formation of lysophosphatidylinositol in platelets stimulated with thrombin or ionophore A23187, *J. Biol. Chem.*, 257, 5196, 1982.

94. Billah, M. M. and Lapetina, E. G., Evidence for multiple metabolic pools of phosphatidylinositol in stimulated platelets, *J. Biol. Chem.*, 257, 11856, 1982.

95. Siess, W., Cuatrecasas, P., and Lapetina, E. G., A role for cycloxygenase products in the formation of phosphatidic acid in stimulated platelets. Differential mechanisms of action of thrombin and collagen, *J. Biol. Chem.*, 258, 4683, 1983.

96. Siess, W. and Lapetina, E. G., Properties and distribution of phosphatidylinositol-specific phospholipase C in human and horse platelets, *Biochim. Biophys. Acta*, 752, 329, 1983.

97. Cheng, H.-L., Blattner, F. R., Fitzmaurice, L., Mushinski, J. F., and Tucker, P. W., Structure of genes for membrane and secreted murine IgD heavy chains, *Nature (London)*, 296, 410, 1982.

98. Sugimoto, Y., Whitman, M., Cantley, L. C., and Eridson, R. L., Evidence that the Rous sarcoma virus transforming gene product phosphorylates phosphatidylinositol and diacylglycerol, *Proc. Natl. Acad. Sci. U.S.A.*, 81, 2117, 1984.

99. Kettman, J. R., Cambier, J. C., Uhr, J. W., Ligler, F., and Vitetta, E. S., The role of receptor IgM and IgD in determining triggering and induction of tolerance in murine B cells, *Immunol. Rev.*, 43, 69, 1979.

100. Cambier, J. C. and Monroe, J. G., B cell activation. V. Differential signaling of B cell membrane depolarization, increased I-A expression, G$_0$ to G$_1$ transition and thymidine uptake by anti-IgM and anti-IgD antibodies, *J. Immunol.*, 133, 576, 1984.

101. Phillips, N. E. and Parker, D. C., Cross-linking of B lymphocyte Fcγ receptors and membrane immunoglobulin inhibits antiimmunoglobulin-induced blastogenesis, *J. Immunol.*, 132, 627, 1984.

102. Garrels, J. I., Two-dimensional gel electrophoresis and computer analysis of proteins synthesized by clonal cell lines, *J. Biol. Chem.*, 254, 7961, 1979.

103. Cambier, J., Unpublished observations.

Chapter 2

REGULATION OF GENE EXPRESSION DURING B CELL DIFFERENTIATION

Dorothy Yuan

TABLE OF CONTENTS

I. INTRODUCTION

Activation of resting B lymphocytes leads to a complex series of events, some of which result in a permanent conversion of the cellular phenotype which clearly distinguishes the differentiated cell from the resting cell. This conversion reflects changes involving regulatory and structural, as well as secretory, proteins. One way by which the regulatory basis of these alterations can be understood is to trace each step of the total biosynthetic pathway, from DNA to RNA, to the final expression of the protein and determine the level at which changes are induced after B cell activation. Therefore, it is the purpose of this review to, first of all, document some of the better-studied protein changes clearly associated with B cell differentiation and then to trace the points in the biosynthetic pathway at which these changes are initiated, starting at the genomic level and ending at the level of protein expression. The most detailed studies involve the immunoglobulin (Ig) proteins. However, it will be seen that due to their unique quality of being able to function both as membrane receptors and as secreted soluble proteins, the analysis of the regulated changes exhibited by these two functional forms of Ig illustrates very well how gene expression can be modulated at a number of molecular levels.

It should be noted that in order to obtain sufficient numbers of cells for biochemical analysis, all of the studies described below involve the use of cells which have been polyclonally stimulated, usually with lipopolysaccharide (LPS). One cannot predict whether all of the events studied in LPS-stimulated cells will be identical to those occurring in B lymphocytes responding to antigen. However, limited biochemical studies of enriched populations of antigen-specific B cells have shown that changes induced by antigen are not significantly different from those observed after LPS stimulation.[1] Therefore, it is likely that most of the conclusions generated from the use of polyclonally stimulated cells can be confirmed in antigen-specific responses if and when studies of the latter become technically feasible.

II. CHANGES IN PROTEIN EXPRESSION DURING B CELL DIFFERENTIATION

For the B lymphocyte, the most prominent change incurred after activation is that of Ig secretion. The initiation of secretion involves not only the increased expression of some well-studied structural proteins, namely, μ and L chains for secreted IgM, and J chain for the initiation of polymerization,[2] but also, the induction of a large number of other regulatory proteins. However, little is known regarding the nature of these latter proteins.

In addition to the induction of Ig secretion, the expression of both of the antigen receptors for the mature, resting B lymphocyte, IgM and IgD, also can be affected by activation. Earlier studies[3] claimed a decrease in cell surface IgM after stimulation of B lymphocytes with LPS. However, the anti-Ig reagent used in these experiments did not distinguish between mIgM and mIgD. More recent cytofluorometric analysis using IgM-specific antisera have failed to show significant differences in mIgM expression after LPS stimulation of normal B lymphocytes.[4] Still, it should be noted that due to the considerable change in size and metabolic activity of B lymphocytes after activation, a small alteration in mIgM expression may be difficult to quantitate. These problems can be circumvented by examining B lymphoma cell lines such as BCL_1[5] and WEHI 231,[6] which can respond to differentiation signals by initiation of Ig secretion without gross changes in cell size. It was found in these cases that mIgM is significantly decreased after stimulation by LPS.[7-9]

A much more dramatic change in membrane expression of IgD is exhibited by normal B lymphocytes upon LPS stimulation. This change was first documented by cell surface iodination[10] and later confirmed by cytofluorometric analysis.[4] After 48 hr of culture with LPS, mIgD of large cells has been estimated to decrease by 80%.

The acquisition of other Ig isotypes by B cells as they further differentiate into memory cells and plasma cells can be studied in cells cultured with mitogens for more extended periods. A substantial fraction of the cells in these cultures bear other surface isotypes,[11] i.e., mIgG and mIgA, at first in combination with mIgM, but later they begin to exhibit Ig class restriction by expressing only one isotype.

Other than mIg, expression on the B cell surface of some of the antigens encoded by the immune response genes (Ia antigens) has also been shown to increase dramatically after activation by either cross-linking of Ig receptors with antiimmunoglobulin reagents or by LPS stimulation.[12,13] However, the molecular basis for the change has not yet been examined.

In addition to Ig and Ia, it is clear that the expression of many other proteins are also altered during B cell differentiation. A dramatic illustration of these protein changes can be seen in the two-dimensional gel electrophoretic (2-D SDS-PAGE) patterns of proteins extracted from B lymphocytes which had been labeled with ^{35}S-methionine at various times after polyclonal stimulation with mitogens, such as LPS and dextran sulfate (DxS). As shown in Figure 1, a large fraction of the total spectrum of labeled proteins is affected.[14] Moreover, the number of alterations revealed by this type of analysis must be an underestimate since large regions of the autoradiograms contain spots that are too dense to be resolved. Furthermore, proteins with turnover rates that are lower than the labeling period might not be detected. Although very little is known regarding either the identity of the multitude of proteins that are altered or the chemical basis for the changes, however, this type of analysis provides one with some idea of the extent of change in gene expression which occurs during B cell differentiation.

The regulation of alterations in expression of a protein can occur at several levels. These may entail modifications of the gene coding for the protein, changes in various aspects of RNA synthesis, and finally, modifications in post-translational processing of completed proteins. The regulatory basis for alterations in gene expression after B cell activation has been studied in the greatest detail for secretory and membrane Igs. The evidence for involvement at each of the levels will be examined below.

III. CHANGES AT THE GENOMIC LEVEL DURING B CELL DIFFERENTIATION

A. DNA Rearrangement

It is now quite clear that the early stages of B cell maturation are accompanied by specific nucleic acid rearrangements at the immunoglobulin heavy- and light-chain gene loci involving the recombination of V_H, D, and J_H gene segments and of V_L and J_L segments, respectively (reviewed by Tonegawa[15]). Since the subject of this review is restricted to changes in gene expression in response to B cell activation, alterations in gene expression prior to the acquisition of the mature B cell phenotype will not be discussed.

Evidence for further gene rearrangement after B cell activation was derived from Southern blot analysis and partial sequencing of the DNA of plasmacytomas[16] and hybridomas.[17] These DNA rearrangements involve the deletion of genes 5′ to the expressed heavy-chain gene by means of recombination at the "switch" sequences located immediately upstream of Cμ, Cγ, Cα, and Cε genes.[16] Recently, Radbruch and Sablitzky[18] were able to confirm the occurrence of such a switch in normal, nontransformed B lymphocytes by analyzing the DNA status of IgG-bearing cells selected from LPS-stimulated cell populations by use of the fluorescence activated cell sorter

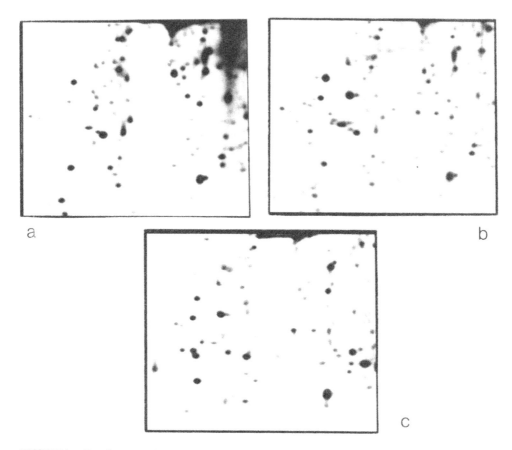

FIGURE 1.   Protein pattern for murine spleen cells labeled with ^{35}S-methionine over the course of the first 5 days after activation with the B cell mitogens LPS + DxS.[14] (a) Labeling on the second day, 2-day film exposure; (b) labeling on the third day, 1-day film exposure; (c) labeling on the fifth day, 1-day film exposure. Cell lysates were subjected first to IEF using an ampholine range of pH 2.5 to 9.5, then to SDS-PAGE with a gradient of 10 to 20% acrylamide. The figure shows a region of the radiofluorograph that was selected to be free of Ig chains. (From Kettman, J. and Lefkovitz, I., *Eur. J. Immunol.*, 14, 778, 1984. With permission.)

(FACS). However, although they showed that μ genes are clearly deleted from cells which bear *only* mIgG, it is not clear whether cells at an earlier stage of differentiation, that is, when they are coexpressing both IgM and IgG, need to have deleted the μ gene as well. The possibility remains that prior to terminal differentiation into plasma cells, the simultaneous production of μ and γ mRNA can be affected by means of splicing of a long transcript. Such a long transcript has been postulated to exist in cells producing both IgM and IgE.[19]

B. Methylation Status of Expressed Genes

In addition to DNA rearrangement, a further regulatory event that can occur at the genomic level involves changes in DNA conformation in localized regions which may be caused by hypomethylation of the gene. The general methylation status of a gene can be estimated by digestion of the DNA with the endonuclease enzyme pair Hpa II and MspI. The sequence CCGG is cleaved by both enzymes, whereas methylation of the inner C renders the sequence CC[m]GG resistant to cleavage by HpaII. A correlation between undermethylation and gene activity has been observed in some systems,[20-22] however, the correlation is by no means universal.[23] After examination of a number of plasmacytomas secreting different Ig classes, Storb and Arp[24] concluded that production of each Ig class is accompanied by changes in the methylation status of the genes involved. Thus, they showed that active transcription of a particular C_H gene is always accompanied by undermethylation of at least some sites within the gene. However, undermethylation is often *not* accompanied by active transcription. For example, the μ-producing myeloma PC3741 does not show rearrangement or detectable transcription[4] of Cγ2b and Cα genes, nevertheless, both of these genes are partially undermethylated.[24]

There have also been some attempts to correlate undermethylation of regions of the Cμ gene with the secretory activity of lymphomas and plasmacytomas.[25] Blackman and Koshland found that all of the CCGG sites within the Cμ exons are hypomethylated in IgM-secreting cell lines (see Figure 2), while these sites remain heavily methylated in less mature cell lines. However, studies in other laboratories have shown that the Cμ gene of an immature cell line, 18—81,[26] which has the characteristics of a nonsecreting pre-B cell line, is already partially undermethylated.[24] Furthermore, all of the CCGG sites within the Cμ gene in a cloned BCL$_1$ in vitro line, which contains mainly mRNA for μ chain of mIgM (μm) and does not secrete Ig without stimulation,[27] has been shown to be hypomethylated.[28] Therefore, the correlation between hypomethylation and secretion cannot be established unless a change in methylation status of the μ gene can be detected in the same cell population after it has been induced to differentiate.

Since no further DNA rearrangement other than the initial VDJ joining is required for the transcription of the Cδ gene, an attempt has also been made to correlate the activation of Cδ with the methylation status of the gene. However, although it was found in one IgD-expressing cell line that part of the δ gene is hypomethylated,[29] subsequent analysis of a number of δ mRNA-producing cell lines showed that the δ gene retains the same methylation status as nonexpressing cell lines[30] (see Figure 3).

A much more satisfactory correlation has been shown for the methylation status of the J chain gene, in that in nonlymphoid and B lymphoma cells not producing J chains (WEHI 231, K46R, and L10A), both CCGG sequences within the gene are resistant to HpaII digestion, whereas in hybrid and plasmacytoma DNAs (WEHI 279.1, M × W 231.1a2, MPC-11, and MOPC 315) which produce copius amounts of J chain, these sequences as well as another site 3′ to the gene are completely susceptible to Hpa II cleavage.[31] These results suggest that the increased J chain production in plasmacytoma cells may be associated with chromatin changes that are propagated through the entire J gene and the flanking sequences resulting in the activation of this gene.

In conclusion, the initial VDJ joining which results in the creation of a functional μ gene is accompanied by well-documented changes in methylation status of the genes involved, as well as increased endonuclease sensitivity of the region.[32] Similar studies investigating the role of DNA conformation in regulating changes in protein expression after B cell differentiation have been conclusive only in the case of the activation of the J chain gene. Here, a strict correlation has been observed between hypomethylation and the presence in the cell line of easily detectable amounts of the J chain, suggesting that this gene can be transcribed efficiently only after hypomethylation. However, the relationship between DNA methylation and the other phenotypic alterations that occur during B cell differentiation is still tenuous.

It should be noted, in addition, that other than the inner cytosine in the sequence CCGG, many other cytosines in DNA are methylated. Therefore, analyses restricted to sensitivity to HpaII digestion may not be adequate to define the methylation status of a gene.

C. Role of RNA Transcription Enhancers and Promoters

Transcriptional enhancers, originally discovered in viral genomes, are short, cis-acting, regulatory sequences that strongly stimulate transcription from promoters of

Cμ1 Cμ2   Cμ3 Cμ4 μs

CELL LINE					Ig Secretion
Liver[24]	+	+	+	+	-
18-81[24]	±	-	±	N.D.	-
WEHl 231[25]	+	+	+	+	-
K46R[25]	+	+	+	+	-
BCL$_1$[28]	-	-	-	-	-
WEHl 279.1[24,25,29]	-	-	-	±	+
MOPC 104E[24,25]	-	-	-	-	+
PC 3741[24]	-	-	-	N.D.	+
MxW 231.1a2[25]	-	-	-	±	+

FIGURE 2. Methylation status of CCGG sequences within the Cμ gene. ↑ indicates locations of CCGC sequences; + indicates resistance to Hpa II digestion.

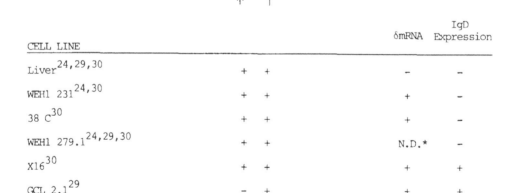

Cδ1 Cδ2 Cδ3   δs   δx   δM1δM2

CELL LINE			δmRNA	IgD Expression
Liver[24,29,30]	+	+	-	-
WEHl 231[24,30]	+	+	+	-
38 C[30]	+	+	+	-
WEHl 279.1[24,29,30]	+	+	N.D.*	-
X16[30]	+	+	+	+
GCL 2.1[29]	-	+	+	+

* N.D. Not determined

FIGURE 3. Methylation status of CCGG sequences within the Cδ gene. ↑ Indicates locations of CCGC sequences; + indicates resistance to Hpa II digestion.

nearby genes (reviewed by Benerji and Schaffner[33]). Recently, such enhancer sequences have been identified both within the J$_H$-Cμ intronic region[34,35] and within the intronic region between Jx and Cx.[36] It is clear from these studies that activation of a specific V region is dependent upon DNA rearrangement which brings it within the influential range of the enhancer. However, since enhancer sequences isolated from both germ line genes and rearranged genes from Ig-secreting cells are equally stimulatory in trans-fection assays, it is not likely that modifications of these sequences are responsible for changes demonstrated by B cells after activation. Rather, they may be the site of action of induced factors which function by increasing the rate of polymerase entry into the promoters of both the heavy- and light-chain genes upon B cell activation (see Section IV.B).

Distinct promoter region structures restricted to the Ig gene family[37] have been described. These consist of the octanucleotide sequences, ATTTGCAT, located some 70 bases upstream of the RNA initiation site in every light-chain gene examined, and the precise inverse of the sequence (ATGCAAAT), occurring in the corresponding location of each heavy-chain gene examined. Therefore, in addition to Ig-specific enhancer sequences, these 5′ promoter sequences may also be sites where factors which modulate the rate of transcription of Ig genes can bind.

IV. CHANGES IN RNA SYNTHESIS DURING B CELL DIFFERENTIATION

A general increase in RNA synthesis, as detected by incorporation of radioactive RNA precursors, has long been associated with B cell differentiation. Recently, a general increase in total RNA content of B lymphocytes shortly after activation has also been documented by means of acridine orange staining followed by flow cytofluorimetric analysis.[38] While a large part of this increase is due to accelerated ribosomal RNA synthesis necessary for the increased metabolic activity of the activated cells, a significant portion of the increase must be correlated with the immense degree of altered expression at the protein level (Figure 1).

A. Changes in Steady-State Levels

Of the complex array of RNAs which must be altered during B cell differentiation, only mRNA for the various Ig proteins has been studied extensively because of the availability of specific DNA probes. The first parameter examined was the steady-state level of each specific mRNA, which is probably the most important regulatory element in determining the translation frequency of a polypeptide. Thus, the commencement of IgM secretion after B cell activation can be correlated with the accumulation of x chain mRNA.[39] More recently, a similar increase (estimated to be on the order of 100-fold in cells cultured for 5 days with LPS) in mRNA of μ chain of secreted IgM (μs) has been shown by Northern blot analysis comparing RNA from nonstimulated and activated cells (Figure 4, panel C).[40] Interestingly, the accumulation of mRNA for J chains has been shown to lag behind the increase in μs mRNA by approximately 24 hr.[25]

Similarly, as shown in Figure 4, panel D, the commencement of IgG$_3$ secretion in LPS-stimulated cells has been shown to be accompanied by an increase in γ3 mRNA.[40] Furthermore, when the cells are stimulated by B cell differentiation factors (BCDFγ), in addition to LPS, the switch to IgG$_1$ production is also preceded by the accumulation of mRNA for γ1 chains[41] (Figure 5).

A similar correlation was observed in the converse case involving the decreased surface expression of IgD in activated B cells. It was found, as shown in Figure 4, panel A, that by 5 days after LPS stimulation, the steady-state concentration of δ mRNA was also decreased dramatically,[40] accounting for the cessation of translation of δ chains.[42]

The change in steady-state levels of mRNA for each protein can be the cumulative result of regulatory events at a number of levels. These include (1) the transcriptional rate, which may be determined by DNA conformation or factors which affect RNA polymerase entry and/or initiation; (2) the processing rate of the initial transcript and the transport rate of the mature mRNA out of the nucleus, which may be determined by capping, splicing, and cleavage as well as poly-A addition enzymes; and (3) the half-

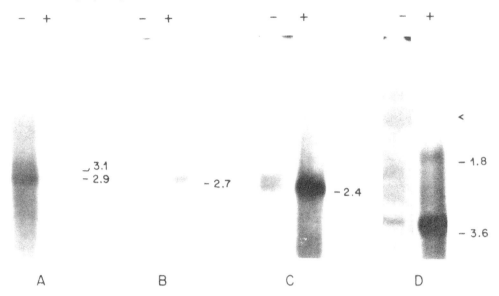

FIGURE 4.   Effect of LPS stimulation on steady-state levels of mRNA for μ, δ, and γ chains.[40] Four fifths of the poly (A)+ RNA (20 μg) from 1.5×10^9 unstimulated splenocytes was loaded in the first lane (−), while the entire yield of poly (A)+ RNA from 2×10^8 day-5 LPS blasts (25 μg) was loaded in the second lane (+). The Northern blot (prepared as in Figure 1) was first hybridized to (A), ^{32}P-labeled δ cDNA, followed by stripping and successive hybridization to (B) μ membrane-specific (μ_m) probe; (C) μ constant region (μ_c) probe; and (D) γ3 constant region (γ) probe. The Northern blots in (A) were exposed for 4 days to allow maximum detection of δ mRNA in LPS-stimulated cells. Because of this, the first lane looks overexposed, while the other blots in this figure were exposed for only 1 day. (From Yuan, D. and Tucker, P. W., *J. Immunol.*, 132, 1561, 1984. With permission.)

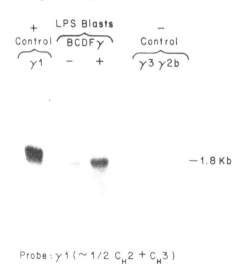

FIGURE 5.   Northern blot analysis of γ_1 mRNA in B cells stimulated with LPS vs. LPS plus PK 7.1 SN containing BCDFγ.[41] The Northern blots were hybridized to a Cγ_1 specific probe. Size of the γ mRNA was calculated from the positions of 18S and 28S rRNA determined by acridine orange staining. (From Jones, S., Chen, Y. W., Isackson, P., Pure, E., Layton, J., Word, C., Krammer, P., Tucker, P., and Vitetta, E. S., *J. Immunol.*, 131, 3049, 1983. With permission.)

life of many mRNAs, which may differ from each other due to the intrinsic RNA structure or the ability to bind to other regulatory elements. Some of these possible regulatory steps have been analyzed in detail for Ig expression and are discussed below.

B. Transcriptional Changes

Because of the transient half-life of nascent RNA and uncertainty regarding the rate of processing of each message, the most accurate determination of relative transcriptional rates of specific genes in vivo is by means of pulse labeling of nascent RNA chains. In order to achieve a high specific activity, isolated nuclei are pulse labeled in vitro. If the labeling period is short relative to the time needed to transcribe each RNA chain, then the extent of hybridization of the labeled RNA to each region of the gene should indicate the number of polymerase arrested (or polymerase loading) in that region at the moment of cell lysis. Darnell and collaborators have shown that their conclusions regarding initiation, transcriptional rate, and termination of RNA derived from in vitro nuclear pulse labeling experiments are similar to those derived from in vivo pulse labeling experiments.[43,44]

Therefore, using this method it is possible to determine if the dramatic increase in μs mRNA levels found in B lymphocytes after activation can be attributed to an increase in the transcriptional rate of the $C\mu$ gene. Thus, Yuan and Tucker[45] found that after LPS stimulation of normal B lymphocytes, the transcriptional level of the $C\mu$ gene is increased 6- to 12-fold by 5 days after activation (Figure 6). The extent of increase was determined by measuring the percent of total labeled RNA hybridized to the $C\mu$ probe for each cell population. Therefore, while this measurement does not take into account the general increase in RNA synthesis exhibited by LPS-stimulated cells, it allows the quantitation of the preferential engagement of polymerases on the $C\mu$ gene after B cell activation. This observation suggests that factors are induced upon activation which increase the rate of polymerase entry and/or initiation of the Ig gene complex. As discussed earlier, these factors could act by binding selectively to either the enhancer or promoter sequences.

The increase in transcription of the $C\mu$ gene does not extend into the C_d gene, suggesting that additional factors must be also induced to cause the increased termination of the transcript 3' to the $C\mu$ gene. Moreover, it can also be seen in Figure 6 that the approximate location of polymerase termination for the μ mRNA is the same for both nonstimulated (making predominantly μm mRNA) and activated (making predominantly μs mRNA) B lymphocytes, i.e., somewhere within the sequences defined by probe no. 3. The same termination site also appears to be used by hybridoma cells which are making essentially only μ chains of secreted IgM (μs).[46] Therefore, the ratio of μs vs. μm production in B lymphocytes is not governed by differential termination, but must be mediated by cleavage enzymes which recognize alternative polyA addition sites.[47]

To recapitulate, then, it appears that to achieve the increased μs mRNA levels in LPS-stimulated cells, the enhanced production of at least three factors or sets of factors operating at the transcriptional level are required: one to increase the polymerase entry and/or initiation frequency, one to preferentially terminate the nascent μ transcript, and, lastly, one to preferentially cleave at the appropriate site for polyadenylation of the μs premRNA. None of these regulatory elements has been defined, and their interrelationships are not known. However, it may be possible to uncouple these events by examination of B cell tumors which, presumably, may be arrested at one or other of these steps. Studies with the B lymphoma cell line BCL$_1$[5] may serve this purpose. The predominant species of μ mRNA in tumor cells isolated from peripheral blood of tumor-bearing mice and in the cloned BCL$_1$[27] in vitro line, both of which secrete insignificant amounts of IgM, is μm mRNA. After stimulation of these cells to

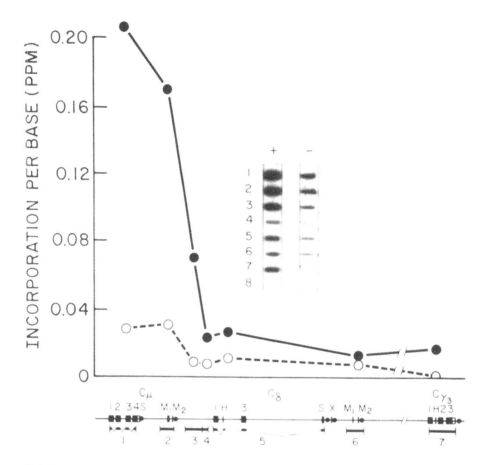

FIGURE 6.   Transcriptional profile of nonstimulated and LPS-stimulated B lymphocytes. Nuclei were prepared from either nonstimulated B lymphocytes (−, ○) or 4-day LPS blasts (+, •) and pulse labeled in vitro with ^{32}P-UTP. RNA was extracted and hybridized to a panel of probes containing DNA inserts as shown in the bottom of the figure. Probe no. 8 consists of the vector pBR 322 with no insert. The incorporation for the DNA region covered by probe no. 1 was determined by direct scintillation counting, while the incorporation for the remaining regions were determined from the relative densities of the bands obtained by densitometer scanning of the autoradiogram. Since the incorporation per base was calculated as the percent total input counts hybridized, the plots should represent the percent of total RNA transcripts engaged in each region of the chromosome at the time of labeling. (Reproduced from *The Journal of Experimental Medicine*, 1984, 160, 564, by copyright permission of The Rockefeller University Press.)

IgM secretion by either LPS[9] or a B cell differentiation factor for IgM secretion, BCDFμ,[27] a significant increase in μs mRNA can be demonstrated which correlates with the ability of these cells to now secrete IgM. However, although some increase in total steady-state levels of both μs and μm mRNA was observed after stimulation (Figure 7), the extent of increase was by no means as great as that observed in LPS-stimulated normal B lymphocytes.

The transcriptional rate of BCL₁ cells has also been examined by in vitro pulse labeling of nascent chains. It was found that the initiation of Ig secretion after LPS stimulation is *not* accompanied by an increase in the transcriptional level of the Cμ gene (calculated as the percent of total input cpm hybridized to Cμ) (Figure 8). Interestingly, the transcriptional level of Cμ in a hybridoma cell line constructed from the BCL₁ tumor (BCL₁xSP2/0), which secretes copious amounts of IgM and makes almost exclusively μs mRNA,[48] was also not significantly higher than nonsecreting lymphoma cells.[45] These findings, when compared to those from normal B lymphocytes (Table 1), suggest that despite the fact that BCL$_1$ lymphoma cells can be induced to Ig secretion, their capacity to differentiate is partially impaired in that they cannot increase the frequency of polymerase initiation and/or entry into the Ig gene complex. In contrast, since the level of Cδ transcription is relatively low in these cells, it is possible that the factor(s) responsible for preferential termination of RNA 3' to Cμ has already been induced, suggesting prior activation of the cells. However, the definite increase in μs mRNA concentration observed in LPS-stimulated cells suggests that the third set of factors, those necessary for the preferential cleavage of the primary transcript into the μs mRNA precursor, is still subject to induction.

The observation that DNA in the region of the J chain gene is completely demethylated[31] only in cells actively producing J chains suggests that the increased level of J chain mRNA may also be mediated by increased transcription.

Possible transcriptional control for the decrease in steady-state mRNA for δ chains has also been explored by using the same method. It was found that the relative transcriptional level of Cμ vs. Cδ in a number of lymphoma cell lines was similar despite great differences in the level of expression of μ vs. δ cytoplasmic mRNA.[30] These results suggested that post-transcriptional events may play a greater role in determining the final steady-state concentration of the δmRNA. Interestingly, the Cδ transcriptional level measured in LPS-stimulated B lymphocytes was also found to be not significantly decreased when compared to that of unstimulated B lymphocytes,[44] suggesting that the regulatory basis for the decreased steady-state concentration of δ mRNA is also post-transcriptional.

However, as activated B lymphocytes become plasma cells, the Cδ transcription may be eventually discontinued. Although plasma cells have not been examined, studies of their presumed transformed counterpart, myeloma[30] or hybridoma[30,46] cells, have shown that Cδ transcription is indeed completely turned off. Therefore, Cδ transcription may provide a better marker for this previously ambiguous step in B cell differentiation. Thus, it is possible that the continuation of δ transcription in an activated cell permits the polymerase to proceed through to further downstream genes such as Cγ, resulting in the synthesis of γ mRNA without gene deletion. However, once the cell has terminally differentiated into plasma cells, this option is then eliminated.

C. Regulation at the RNA Processing Level

Some clues regarding the type of post-transcriptional regulation which may be operating to decrease the level of δmRNA may be derived from short-term in vivo labeling experiments, in which it was found that the amount of newly synthesized cytoplasmic δ mRNA is much decreased in LPS-stimulated cells when compared to resting B cells.[40] Therefore, processing events in the nucleus must play an important regulatory role in determining the amount of δ mRNA which finally appears in the cytoplasm. The mode of regulation of δ mRNA expression in activated B lymphocytes may be different, therefore, from that in resting B lymphocytes, in which the half-life of cytoplasmic δ mRNA may play a more important role. In the latter case a higher amount of short-term labeled δ mRNA was found in the cytoplasm, although the steady-state level as determined by Northern blot analysis was low.[40]

The exact nature of regulatory events occurring at the nuclear level is still unclear. However, comparison of the abundance of nuclear RNA species with that of cytoplasmic δ mRNA in various lymphoma cell lines led Mather et al.[30] to conclude that differential nuclear RNA half-life and/or differences in efficiency of splicing may be involved in different lymphoma cells. It will be of interest to determine which of the above alternatives is involved in the case of normal activated B cells.

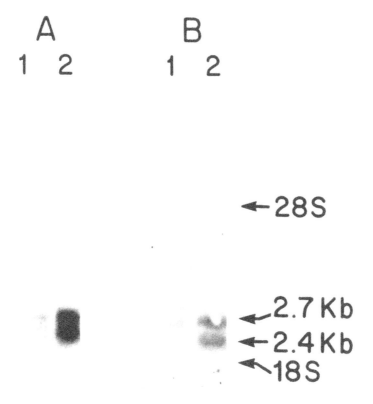

FIGURE 7.   The effect of LPS stimulation on amounts of RNA specific for the μs and μm.[9] PBLs from BCL$_1$ tumor-bearing mice were cultured for 3 (A) or 5 (B) days in the presence (2) or absence (1) of LPS; 60 μg of RNA was loaded in channel A1 and A2, while 25 μg of RNA was loaded in channels B1 and B2. Positions of 28S and 18S ribosomal RNA were determined by staining gels with ethydium bromide. RNA separated by agarose gel electrophoresis was transferred to nitrocellulose and hybridized with ^{32}P-labeled μ-specific probe. (Reproduced from *The Journal of Experimental Medicine*, 1984, 160, 564, by copyright permission of The Rockefeller University Press.)

D. Regulation of mRNA Half-Life

Although no significant enhancement in the transcriptional level of μ mRNA is detected after stimulation of BCL$_1$ cells, it is clear that there is a considerable increase in the total amount of both μs and μm mRNA, as detected by Northern blot analysis (see Figure 7). Although not easily proved, this increase in mRNA concentration without a concomitant increase in transcription can be most easily attributed to a specific increase in RNA stability. Furthermore, if these conclusions can be extrapolated to normal B lymphocytes, it is possible that an increase in mRNA half-life, in addition to the increase in transcription levels, may both contribute to the increased levels of μs mRNA found in LPS-stimulated cells.

More quantitative calculations of the accumulation of mRNA for γ chains in the myeloma cell line MPC 11[49] have resulted in similar conclusions. Likewise, the increase in J chain mRNA after LPS stimulation has been attributed to both an increase in transcriptional levels as well as increased mRNA stability.[25] The nature of the factors responsible for the increased stability remains unknown.

V. REGULATION AT THE PROTEIN LEVEL

So far it can be seen that much of the alterations in Ig secretion and mIg expression

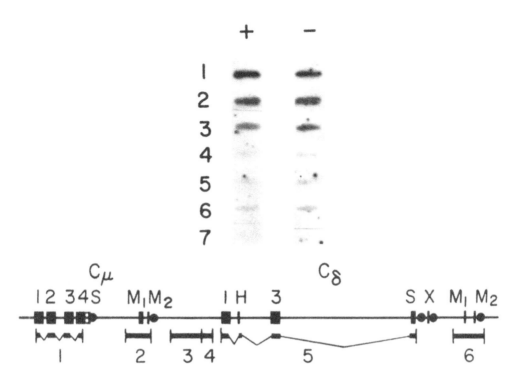

FIGURE 8.   Effect of LPS stimulation on the transcription of BCL₁ cells. BCL₁ cells from the peripheral blood lymphocytes of tumor-bearing mice were cultured with LPS for 3 days. Nuclei prepared from an equal number of stimulated (+) or nonstimulated (−) cells were pulse labeled in vitro with ^{32}P-UTP. Extracted RNA was hybridized to the panel of probes containing DNA inserts as shown in the bottom of the figure. Probe no. 7 consists of the vector pBR 322 with no insert. Separate aliquots of the same cells were labeled with ^{35}S-methionine and immunoprecipitated with anti-μ antibodies to insure that the LPS-stimulated cells were secreting IgM.

Table 1

COMPARATIVE EFFECTS OF LPS STIMULATION ON NORMAL B
LYMPHOCYTES AND BCL₁ CELL LINES

	B cells[40,45]		BCL₁ in vivo[49]		BCL₁5B in vitro[8]	
	No LPS	+LPS	No LPS	+LPS	No LPS	+LPS
Enhanced polymerase initiation	−	+	−	−	N.D.[a]	N.D.
Enhanced termination 3′ to μm	−	+	+	+	N.D.	N.D.
Increased μs RNA cleavage	−	+	−	+	N.D.	N.D.
High steady-state level of μm RNA	−	+	−	+	+	+
Cytoplasmic IgM	−	+	−	+	+	+
IgM secretion	−	+	−	+	−	+

[a]   N.D. = not determined.

occurring after B cell differentiation can be attributed to regulated changes at the level of RNA synthesis. However, although the translation rate of a particular polypeptide usually reflects the steady-state concentration of its message, the newly synthesized polypeptide chain is subjected to further modifications as it moves through the intracellular compartments before it is finally expressed on the cell surface. Detailed studies of both μm and μs[50-53] have shown, for example, that after being synthesized on membrane-bound polyribosomes, both chains are initially glycosylated by the *en bloc* addition of core sugars. Further trimming of the carbohydrate moieties of the two chains and association with light chains occur as the polypeptides proceed through the endoplasmic reticulum. Finally, terminal sugars are added in the Golgi cisternae to produce the mature cell surface and secreted proteins. The biosynthesis of IgD has been shown to follow a similar pathway as IgM.[54-56] An incompletely glycosylated cytoplasmic precursor is processed in the endoplasmic reticulum and terminally glycosylated shortly before surface expression. Recent experiments[42] have shown that the intracellular transit time, half-life of the nascent μm, and the half-life of cell surface IgM are considerably different from that of nascent δm and cell surface IgD, suggesting that these post-translational events are under stringent regulation. Analysis of BCL$_1$ cells which have been stimulated with LPS suggests that additional modifications may be superimposed on these regulated events after activation.[9] Here it was found that despite the higher concentration of cytoplasmic μm mRNA in LPS-stimulated cells, the rate of accumulation of cell surface IgM is 50% that of the rate in nonstimulated tumor cells. Since both μm and μs mRNA can bind to polyribosomes equally well, it is likely that the site of regulation is post-translational. Moreover, since the half-life of cell surface IgM in the LPS-stimulated cells is not lower than that of nonstimulated cells, the post-translational event which is modified must occur during the intracellular transit of the polypeptide chain. This may involve, for example, the availability of glycosylation enzymes or a differential ability to bind to light chains.

Another case of post-translational regulation has been documented in the in vitro line of BCL$_1$ (BCL$_1$5B)[8] cells which secrete a higher basal level of IgM than the line cloned by Brooks et al.[27] It was found that the effects of LPS stimulation on BCL$_1$ 5B cells differs from that of the in vivo line in that there is no effect on RNA synthesis (see Table 1). In contrast, LPS stimulation results in the increased secretion of preformed Ig already present in the cytoplasm before stimulation. Interestingly, the decrease in surface IgM after LPS stimulation observed in the in vivo line of BCL$_1$ cells also occurs in BCL$_1$5B, suggesting that mIgM expression in cells which have reached a relatively late stage in the differentiation pathway can be further modulated by LPS stimulation. It is possible, therefore, that in normal B lymphocytes, expression of mIgM only decreases near the final stages of differentiation.

VI. CONCLUSIONS

This review has shown that the alterations of Ig expression during B cell differentiation can result from changes at virtually all levels of gene expression. At the genomic level, further gene rearrangements subsequent to the initial VDJ joining occur to allow the high rate secretion of Ig isotypes other than IgM and IgD; although methylation changes of at least one gene (J chain gene) are definitely necessary for the expression of pentameric IgM, further modification of other genes may also be involved. The most important changes which result in the initiation of Ig secretion occur, however, at the transcriptional level. Activation of the B lymphocyte results in the enhanced initiation of transcription of the Ig gene complex, the increased termination of the polymerase 3' to the μ gene, and increased cleavage of the preRNA at the appropriate site for μs mRNA production. All of these events culminate in a greater production rate of μs mRNA and, together with a possible extension in μ mRNA half-life, account for the enhanced accumulation of μs mRNA in the cytoplasm. The steady-state level of mRNA for J chains is also increased after LPS stimulation, resulting in the synthesis of large amounts of secretory IgM which is exported from the cell.

In contrast, the changes in mIg during B cell differentiation seem to be more a result of changes at post-transcriptional levels. Although the decrease in mIgD is correlated with a marked decrease in the cytoplasmic δ mRNA concentration, it does not appear to be a direct result of changes at the transcriptional level. Rather, alterations in the processing or half-life of the nuclear δ mRNA are implicated. It is intriguing that the decrease in mIgM may, on the other hand, be regulated at a yet more distal level, involving translational or post-translational alterations.

Now that the exact levels of regulation have been documented for the changes in IgM and IgD expression during B cell differentiation, the basis has been laid for further probes into the mediators responsible for these changes. The most likely candidates responsible for many of these changes are factors which are induced after B cell activation, and it is likely that they are contained within the altered species observed in the 2-D SDS-PAGE analysis of total B cell proteins after activation (Figure 1).

Finally, in addition to the data for normal cells, some detail has been presented in this review showing the molecular changes occurring in the BCL₁ cell line during its differentiation. It is apparent that while many parallels can be drawn between this transformed cell line and normal cells, there are also differences. It is not clear at present whether the differences arise from aberrations due to the neoplastic nature of BCL₁ cells or whether this tumor represents B cells at an arrested stage of differentiation; this fortunately can still respond to further differentiation signals. Yet another explanation is that while the majority of B lymphocytes are activated by LPS, the response is undoubtedly heterogenous, so that at any given time point after stimulation the population can contain cells which belong to a number of differentiation pathways. The response of BCL₁ cells to LPS, on the other hand, may be restricted to only one of these pathways. Therefore, until methods become available for the definitive separation of subpopulations of activated B cells, tumor cells will continue to be of value in the dissection of the molecular control of B cell differentiation.

ACKNOWLEDGMENTS

I thank Ms. Tam Dang for expert technical assistance and Ms. Laura Broyles for cheerful secretarial assistance. I am indebted to Dr. Phil Tucker for many stimulating discussions and suggestions regarding the manuscript. I am also indebted to Dr. K. Brooks, Dr. J. R. Kettman, Ms. E. A. Weiss, and Mr. W. A. Kuziel for critically reviewing the manuscript.

REFERENCES

1. Kanowith-Klein, S., Vitetta, E. S., Kern, E. L., and Ashman, R. F., Antigen-induced changes in the proportion of antigen-binding cells expressing IgM, IgG and IgD receptor, *J. Immunol.*, 122, 2349, 1979.
2. Koshland, N. Z., Structure and function of the J chain, *Adv. Immunol.*, 20, 41, 1975.
3. Melchers, F. and Andersson, J., The kinetics of proliferation and maturation of mitogen-activated bone marrow-derived lymphocytes, *Eur. J. Immunol.*, 4, 687, 1974.
4. Monroe, J. G., Havran, W. L., and Cambier, J. C., B lymphocyte activation: entry into cell cycle is accompanied by decreased expression of IgD but not IgM, *Eur. J. Immunol.*, 13, 208, 1983.
5. Slavin, S. and Strober, S., Spontaneous murine B cell leukemia, *Nature (London)*, 272, 624, 1977.
6. Warner, N. L., Harris, A. W., McKenzie, I. F. C., De Luca, D., and Gutman, G., Lymphocyte differentiation as analyzed by the expression of defined cell surface markers, in *Membrane Receptors of Lymphocytes*, Seligmann, M., Preud'homme, J. L., and Kourilsky, F. M., Eds., Elsevier, New York, 1975, 203.
7. Boyd, A. W., Goding, J. W., and Schrader, J. W., The regulation of growth and differentiation of a murine B cell lymphoma. I. Lipopolysaccharide-induced differentiation, *J. Immunol.*, 126, 2461, 1981.
8. Lafrenz, D., Koretz, S., Stratte, P. T., Werd, R. B., and Strober, J., LPS-induced differentiation of a murine B cell leukemia (BCL): changes in surface and secreted IgM, *J. Immunol.*, 129, 1329, 1982.

9. Yuan, D. and Tucker, P. W., Effect of LPS stimulation on the transcription and translation of mRNA for cell surface IgM, *J. Exp. Med.*, 156, 962, 1982.
10. Bourgois, A., Kitajima, K., Hunter, I. R., and Askonas, B. A., Surface immunoglobulins of lipo-polysaccharide-stimulated cells. The behavior of IgM, IgD, and IgG, *Eur. J. Immunol.*, 7, 151, 1977.
11. Kearney, J. F., Cooper, M. D., and Lawton, A. R., B lymphocyte differentiation induced by lipo-polysaccharide. IV. Development of immunoglobulin class restriction in precursors of IgG-synthes-izing cells, *J. Immunol.*, 117, 1567, 1976.
12. Mond, J. J., Seghal, E., Kung, J., and Finkelman, F. D., Increased expression of I-region-associated antigen (Ia) on B cells after crosslinking of surface immunoglobulin, *J. Immunol.*, 127, 881, 1981.
13. Monroe, J. G. and Cambier, J. C., Levels of mIa expression on mitogen-stimulated murine B cells is dependent on position in cell cycle, *J. Immunol.*, 130, 626, 1983.
14. Kettman, J. and Lefkovits, I., Changes in the protein pattern of murine B cells after mitogenic stimulation, *Eur. J. Immunol.*, in press.
15. Tonegawa, S., Somatic generation of antibody diversity, *Nature (London)*, 302, 375, 1983.
16. Honjo, T., Immunoglobulin genes, *Ann. Rev. Immunol.*, 1, 499, 1983.
17. Hurwitz, J. L., Coleclough, C., and Cebra, J. J., C_H gene rearrangements in IgM-bearing B cells and in the normal splenic DNA component of hybridomas making different isotypes of antibody, *Cell*, 22, 349, 1980.
18. Radbruch, A. and Sablitzky, F., Deletions of $C\mu$ genes in mouse B lymphocytes upon stimulation with LPS, *EMBO J.*, 11, 1929, 1983.
19. Yaoita, Y., Kumagai, Y., Okumura, K., and Honjo, T., Expression of lymphocyte surface IgE does not require switch recombination, *Nature (London)*, 287, 697, 1982.
20. Razin, A. and Riggs, A. D., DNA methylation and gene function, *Science*, 210, 604, 1980.
21. Bird, A., Taggart, M. H., and MacLeod, D., Loss of rDNA methylation accompanies the onset of ribosomal gene activity in early development of *Xenopus laevis*, *Cell*, 26, 381, 1981.
22. Sutter, D. and Doerfler, W., Methylation of integrated adenovirus type II DNA sequences in trans-formed cells is inversely correlated with viral gene expression, *Proc. Natl. Acad. Sci. U.S.A.*, 77, 253, 1980.
23. Busslinger, M., Hurst, J., and Flavell, R. A., DNA methylation and the regulation of globin gene expression, *Cell*, 34, 197, 1983.
24. Storb, U. and Arp, B., Methylation patterns of immunoglobulin genes in lymphoid cells: correlation of expression and differentiation with undermethylation, *Proc. Natl. Acad. Sci. U.S.A.*, 80, 6642, 1983.
25. Koshland, M. E., Presidential address: molecular aspects of B cell differentiation, *J. Immunol.*, 131, i, 1983.
26. Riley, S. C., Brock, E. J., and Kuehl, W. M., Induction of L chain expression in a pre-B cell line by fusion to myeloma cells, *Nature (London)*, 289, 804, 1981.
27. Brooks, K., Yuan, D., Uhr, J. W., Krammer, P. H., and Vitetta, E. S., Lymphokine-induced IgM secretion by clones of neoplastic B cells, *Nature (London)*, 302, 825, 1983.
28. Brooks, K. D., Unpublished observations.
29. Rogers, J. and Wall, R., Immunoglobulin heavy chain genes: demethylation accompanies class switching, *Proc. Natl. Acad. Sci. U.S.A.*, 78, 7497, 1981.
30. Mather, E. L., Nelson, K. J., Haimovich, J., and Perry, R. P., Mode of regulation of immunoglob-ulin μ and δ-chain expression varies during B-lymphocyte maturation, *Cell*, 36, 329, 1984.
31. Yagi, M. and Koshland, M. E., Expression of the J chain gene during B cell differentiation is inversely correlated with DNA methylation, *Proc. Natl. Acad. Sci. U.S.A.*, 78, 4907, 1981.
32. Mather, E. L. and Perry, R. P., Methylation status and DNase I sensitivity of immunoglobulin genes: changes associated with rearrangement, *Proc. Natl. Acad. Sci. U.S.A.*, 80, 4689, 1983.
33. Benerji, J. and Schaffner, W., Transient expression of cloned genes in mammalian cells, in *Genetic Engineering — Principles and Methods*, Vol. 5, Stelow, J. K. and Hollaender, A., Eds., Plenum Press, New York, in press.
34. Gillies, S. D., Morrison, S. L., Oi, V. T., and Tonegawa, S., A tissue-specific transcription enhancer element is located in the major intron of a rearranged immunoglobulin heavy chain gene, *Cell*, 33, 717, 1983.
35. Banerji, J., Olson, L., and Schaffner, W., A lymphocyte-specific cellular enhancer is located down-stream of the joining region in immunoglobulin heavy chain genes, *Cell*, 33, 729, 1983.
36. Queen, C. and Baltimore, D., Immunoglobulin gene transcription is activated by downstream se-quence elements, *Cell*, 33, 741, 1983.
37. Parslow, T. G., Blair, D. L., Murphy, W. J., and Granner, D. K., Structure of the 5′ ends of immunoglobulin genes: a novel conserved sequence, *Proc. Natl. Acad. Sci. U.S.A.*, 81, 2650, 1984.
38. Darzynkiewicz, Z., Traganos, F., Sharpless, T., and Melamed, M. R., Lymphocyte stimulation: a rapid multiparameter analysis, *Proc. Natl. Acad. Sci. U.S.A.*, 73, 2881, 1976.

39. Tsuda, M., Honjo, T., Shimizu, A., Mizuno, D., and Natori, S., Induced synthesis of immunoglobulin messenger RNA accompanies induction of immunoglobulin production in cultured mouse spleen cells, *Biochem. J.*, 84, 1285, 1978.
40. Yuan, D. and Tucker, P. W., Regulation of IgM and IgD synthesis in B lymphocytes. 1. Changes in biosynthesis of mRNA for μ and δ chains, *J. Immunol.*, 132, 1561, 1984.
41. Jones, S., Chen, Y. W., Isakson, P., Pure, E., Layton, J., Word, C., Krammer, P., Tucker, P., and Vitetta, E. S., Effect of T cell-derived lymphokines containing B cell differentiation factor(s) for IgG (BCDFγ) on γ-specific mRNA in murine B cells, *J. Immunol.*, 131, 3049, 1983.
42. Yuan, D., Regulation of IgM and IgD synthesis in B lymphocytes. II. Translational and post-translational events, *J. Immunol.*, 132, 1566, 1984.
43. Weber, J., Jelinek, W., and Darnell, J. E., Jr., The definition of a large viral transcription unit late in Ad2 infection of HeLa cells: mapping of nascent RNA molecules labeled in isolated nuclei, *Cell*, 10, 611, 1977.
44. Hofer, E. and Darnell, J. E., Jr., The primary transcription unit of the mouse β-major globin gene, *Cell*, 23, 585, 1981.
45. Yuan, D. and Tucker, P. W., Transcriptional regulation of the heavy chain locus in normal murine B lymphocytes, *J. Exp. Med.*, 160, 564, 1984.
46. Yuan, D., Gillium, A. C., and Tucker, P. W., IgM and IgD expression in murine B cells is controlled at multiple stages, *Fed. Proc. Fed. Am. Soc. Exp. Biol.*, 44, 2652, 1985.
47. Blattner, F. and Tucker, P. W., The molecular biology of IgD, *Nature (London)*, 307, 417, 1984.
48. Yuan, D., Unpublished observations.
49. Schibler, U., Marcu, K. B., and Perry, R. P., The synthesis and processing of the messenger RNAs specifying heavy and light chain immunoglobulins in MPC-11 cells, *Cell*, 15, 1495, 1978.
50. Vitetta, E. S. and Uhr, J. W., Cell surface Ig. IX. A new method for the study of synthesis, intracellular transport and exteriorization in murine splenocytes, *J. Exp. Med.*, 139, 1599, 1974.
51. Andersson, J., Lafleur, L., and Melchers, F., IgM in bone marrow-derived lymphocytes. Synthesis, surface deposition, turnover and carbohydrate composition in unstimulated mouse B cells, *Eur. J. Immunol.*, 4, 170, 1974.
52. Vassalli, P., Tarfakoff, A., Pink, J. K. L., and Jafon, J.-C., Biosynthesis of two forms of IgM heavy chains by normal mouse lymphocytes, *J. Biol. Chem.*, 255, 11822, 1975.
53. Sidman, C., B lymphocyte differentiation and the control of IgM μ chain expression, *Cell*, 23, 379, 1981.
54. Goding, J. W. and Herzenberg, L. A., Biosynthesis of lymphocyte surface IgD in the mouse, *J. Immunol.*, 124, 2540, 1980.
55. Finkelman, F. D., Kessler, S. W., Mushinski, J. F., and Potter, M., IgD-secreting murine plasmacytomas: identification and partial characterization of two IgD myeloma proteins, *J. Immunol.*, 126, 680, 1981.
56. Vasilov, R. G. and Ploegh, H. L., Biosynthesis of murine immunoglobulin D: heterogeneity of glycosylation, *Eur. J. Immunol.*, 130, 619, 1983.

Chapter 3

ANTIIMMUNOGLOBULIN ANTIBODY INDUCTION OF B LYMPHOCYTE ACTIVATION AND DIFFERENTIATION IN VIVO AND IN VITRO

Fred D. Finkelman, James J. Mond, and Eleanor S. Metcalf

TABLE OF CONTENTS

I. INTRODUCTION

An essential characteristic of an immune response is its specificity for the immunizing antigen. The discoveries that B lymphocytes bear surface immunoglobulin (sIg)[1-3] and that the sIg of a B cell clone has the same antibody specificity as Ig secreted by activated cells of that clone[4-6] led to rapid acceptance of the concept that B cell sIg plays an important role in antigen-induced activation of antigen-specific B lymphocytes. The question, however, of whether an interaction between B cell sIg and antigen contributes to B cell activation by directly activating B cells,[7] by focusing an antigen-associated or antigen-specific stimulatory signal onto B cells,[8] or by some mixture of these two processes is still not satisfactorily answered. Indeed, attempts to answer this question have been complicated by discoveries that (1) most mature B lymphocytes bear two sIg isotypes, IgM and IgD,[9-15] that on a single cell or cell clone, have identical antigen binding specificity and idiotypic determinants;[4,6,16] and (2) different B cell subsets have different requirements for activation.[17-22] Many in vivo and in vitro studies have been performed in which anti-Ig antibodies were used to characterize the direct and indirect effects of sIg-ligand interactions on B cell activation. Results of these studies demonstrate that (1) sIg-ligand interactions can have a direct activating effect on most B cells;[1,18,23-29] (2) this direct effect can induce B cell proliferation but never induces Ig secretion;[1,18,23-26,28,29] and (3) sIg can participate in the focusing of activating cells or substances onto B cells that can trigger both proliferation and antibody secretion.[8,30-35] This paper will review studies performed with anti-Ig antibodies that lead to these conclusions and outline the present state of knowledge about (1) the relative direct and indirect roles of sIg-ligand interactions in the generation of an immune response; and (2) differences in the roles played by B cell sIgM and sIgD in B cell activation.

II. WHY USE ANTI-Ig ANTIBODIES TO STUDY B CELL ACTIVATION?

The use of anti-Ig antibodies in studies of B cell activation is predicated on the belief that the interaction between anti-Ig antibody and B cell sIg has similar consequences on a polyclonal level to the interaction between an antigen and the sIg of a B cell specific for that antigen. There are several reasons why this need not be true. First, antigen and anti-Ig antibody bind to different sites on the B cell sIg molecule; antigen will be bound by the antigen binding site that is included within the NH_2-terminal domains of sIg heavy and light chains, while anti-Ig antibodies generally bind to sites on Ig heavy- and/or light-chain constant regions. Second, unless monoclonal anti-Ig antibodies are used, anti-Ig antibodies will usually bind to several sites on sIg molecules. This may allow anti-Ig antibodies to cross-link sIg more tightly than would antigen, which binds to a single site on each sIg half-molecule. Third, while IgG anti-Ig antibodies, which are used in most studies, only have two antigen binding sites and thus can bind to no more than two sIg molecules, a single antigen molecule may have multiple representations of a single antigenic determinant. Thus, some antigens may bind to more sIg molecules and more tightly cross-link sIg than can anti-Ig antibodies. Fourth, many anti-Ig antibodies, unlike most antigens, can fix complement and bind to the B cell IgGFc receptor.

While these differences between antigens and anti-Ig antibodies could affect the results of studies of B cell activation performed with anti-Ig antibodies and must be considered in the interpretation of such studies, we feel that the use of anti-Ig antibodies in studies of B cell activation has advantages that outweigh these potential disadvantages. First, the capacity of anti-Ig antibodies to polyclonally interact with B cell sIg eliminates the necessity of purifying antigen-specific B cells for studies of B cell activation. Antigen-specific B cell purification, in addition to being laborious and expensive, can itself distort analyses of B cell activation, since the isolation process itself may activate B çells or inhibit activation.[36] Furthermore, there is no guarantee that the B cells isolated by a technique that requires their binding to antigen have the same characteristics as the entire population of B cells specific for that antigen (e.g., B cells with the highest sIg density, antigen affinity, or inherent stickiness may be preferentially isolated). In addition, studies aimed at identifying possible functional differences for various sIg isotypes can only be performed with isotype-specific anti-Ig antibodies, since an antigen molecule will bind to all sIg isotypes on B cells specific for that antigen. Finally, studies of the relationship between affinity of sIg-ligand interactions and the extent of B cell activation are most readily performed with panels of monoclonal anti-Ig antibodies that have characterized avidities for B cell sIg.

III. EVIDENCE THAT B CELLS CAN BE ACTIVATED BY sIg-LIGAND INTERACTIONS

In vitro and in vivo studies with anti-Ig antibodies have demonstrated that these agents can rapidly induce changes in B cell surface receptor expression and stimulate these cells to enter the cell cycle and to synthesize DNA. Such studies have been performed in vitro with B cells purified to the highest degree possible with existing purification techniques[2-9] and with monoclonal anti-Ig antibodies,[28,37,38] as well as with conventional affinity purified anti-Ig antibodies.[1,18,23-26,28,29]

A. Changes in Expression of Surface Receptors

By 4 hr after the initial inactivation of B cells with anti-μ or anti-δ antibodies small increases can be seen in B cell expression of sIa antigen. Peak B cell sIa expression is detected 24 to 48 hr after the initiation of culture with anti-Ig antibody.[27] A large majority of cultured B cells exhibit this increase in sIa expression, which is not dependent upon a concomitant increase in B cell size and which is seen when cells are cultured with as little as 0.1 to 1.0 μg/ml of anti-Ig antibody.[27,28,39] In vitro culture of B cells with anti-Ig antibodies also stimulates an increase in the expression of a receptor for the lymphokine IL-2 (detected by immunofluorescence staining with a monoclonal anti-IL-2 receptor antibody[139]). In vivo studies in which mice are injected with anti-δ antibodies have demonstrated rapid induction of a receptor for the iron transporting protein transferrin (detected by immunofluorescence staining with an antitransferrin receptor antibody)[40] as well as a receptor for a B cell differentiation factor produced by Concanavalin A-stimulated T lymphocytes (Con A TRF) (detected by absorption experiments).[41] The expression of some membrane proteins, such as the class I histocompatibility antigens, is unchanged in anti-Ig-stimulated cells,[27] while the expression of other surface molecules, such as a receptor bound by the monoclonal antibody Mel 14, which allows lymphocytes to bind to high venous endothelium so that they can home to lymph node and spleen,[42] is greatly decreased when B cells are stimulated by anti-Ig antibodies.[140] Thus, anti-Ig antibodies stimulate B lymphocytes to increase their expression of receptors that promote their interaction with T lymphocytes, with cytokines that promote growth and/or differentiation, and with nutrients required for cell growth and to decrease their expression of a receptor that would permit them to migrate from the site of activation. These effects of anti-Ig antibodies on B cell activation do not, however, appear to be unique; both LPS[141] and at least one T cell-produced cytokine[43,44] also stimulate B cells to express increased sIa, at least under in vitro conditions.

B. Stimulation of Increases in B Cell Size, RNA Synthesis, and DNA Synthesis by Anti-Ig Antibodies

B lymphocytes can be stimulated by polyclonal, affinity purified, or monoclonal antibodies to μ, δ, or light chains under in vitro or in vivo conditions to increase in size and RNA content and to synthesize DNA and divide.[1,18,23-29,37,39] These observations have been made in vitro with both unfractionated spleen cell cultures and with cultures of "highly purified" small B cells. Under optimal conditions a large majority of B cells increase in size and as many as 50% are stimulated to synthesize DNA.[29] In vivo, a large majority of B cells of both normal and nude mice injected with an optimal dose of affinity purified goat antimouse δ (GaMδ) antibody are increased in size 48 hr after injection and a large percentage of these cells synthesize DNA.[28,45] While these observations might suggest that ligand-sIg interactions have an important role in stimulating B cell clonal expansion, the role of antigen-sIg interactions in the physiological stimulation of B cell DNA synthesis remains controversial. Observations concerning the concentrations of anti-Ig antibodies and the concentrations of purified B cells required for the induction of DNA synthesis, as well as differences in the abilities of anti-Ig antibodies and T-dependent antigens to induce B cell proliferation, are at the heart of this dispute and will be discussed individually.

1. Anti-Ig Antibody Concentration

Relatively high concentrations of soluble anti-Ig antibodies must be used to induce B cells to synthesize DNA. Even the most potent anti-Ig antibodies do not stimulate B cell proliferation in vitro at concentrations below 1 to 5 μg/mℓ and optimal stimulation of DNA synthesis is observed at concentrations ten times greater.[24,39,46] In contrast, large increases in sIa expression, as well as modulation of most sIg from B cells, are clearly seen at anti-Ig concentrations below 0.5 μg/mℓ. Similar findings have been made in vivo; most sIgD is modulated from B cells when BALB/c mice are injected with as little as 12.5 μg/mℓ of GaMδ and large increases in sIa expression are seen with injection of 50 μg of this antibody, yet induction of increases in B cell size and DNA synthesis usually requires injection of 100 to 200 μg of GaMδ and peak stimulation of DNA synthesis is seen when 800 μg of GaMδ antibody is injected.[47]

These observations require that something other than, or in addition to, modulation of sIg from the B cell surface is required for entry into the G_1 stage of the cell cycle and the induction of DNA synthesis. Several observations suggest that the requirement for a high concentration of anti-Ig antibodies to induce B cell mitogenesis reflects a need to repeatedly cross-link sIg molecules on the surface of these cells. First, B cell activation by anti-Ig antibodies clearly requires the cross-linking of sIg, since whole IgG anti-Ig antibody molecules and their divalent F(ab')$_2$ fragments will activate lymphocytes, while the univalent Fab or Fab' fragments of these antibodies fail to do so.[26,48] Secondly, in vitro culture of B cells with anti-Ig antibodies for the minimal period of time required to modulate their sIg, followed by washing and further incubation in the absence of anti-Ig antibodies, has little effect on B cell activation. The induction of B cell DNA synthesis requires that these cells be cultured with anti-Ig antibodies for 36 to 48 hr.[29] Cells cultured with anti-Ig antibodies for up to 24 hr show submaximal increases in size but no increased DNA synthesis. The size of these cells will remain stable when they are cultured for a few hours in the absence of anti-Ig antibody and will again start to increase when anti-Ig antibody is added to the culture, until cells undergo mitosis.[29,49] Thirdly, the concentration of anti-Ig antibody required for the induction of DNA synthesis increases with the duration of culture; B cells can be induced to proliferate if incubated with a low concentration of anti-Ig antibody for 24 hr and a high concentration of anti-Ig antibody for an additional 24 hr, but not if incubated with the low concentration for the entire 48-hr period or if incubated first with the high concentration and later with the low concentration.[142] Finally, B cells can be induced to proliferate when cultured with even low quantities of anti-Ig antibody, provided that this antibody is immobilized onto a solid substrate such as agarose or polyacrylamide beads.[23,50,51] These anti-Ig beads are more effective than soluble anti-Ig antibodies or their F(ab')₂ fragments at inducing B cells to proliferate, but are much less effective at modulating sIg from the B cell surface.[52,53] Preincubation of B cells with soluble anti-Ig antibodies greatly reduces the ability of these cells to be stimulated to proliferate when subsequently incubated with particle-bound anti-Ig antibodies.[54] These observations are all compatible with the view that the rate of cross-linking of B cell sIg by ligand must exceed a critical level for a critical period of time to stimulate B cells to synthesize DNA, and that once Ig has been modulated from the B cell surface its rate of replacement is sufficiently slow that higher concentrations of anti-Ig are required to keep the rate of cross-linking of sIg above the critical level.

Regardless of whether this explanation of the need for high concentrations of anti-Ig antibody for the induction of B cell proliferation is correct, the observation that in vitro and in vivo humoral immune responses, which are dependent upon B cell proliferation, can be generated by quantities of T-dependent antigens much lower than the quantities of anti-Ig antibody required to directly induce B cell proliferation suggests that B cell proliferation in T-dependent immune responses can be induced by mechanisms not wholly dependent on a direct sIg-ligand interaction.

2. B Cell Concentration

The ability of anti-Ig antibodies to induce DNA synthesis by "purified" B cells cultured in vitro at relatively high concentrations (2 to 4×10^5 per microtiter plate well) is consistent with the possibilities that induction of B cell DNA synthesis (1) results directly and solely from sIg cross-linking; (2) requires direct contact between B cells that have been bound to each other by anti-Ig antibodies; (3) requires, in addition to sIg cross-linking, the production of stimulatory lymphokines by non-B cells that contaminate the "purified" B cell populations; or (4) requires stimulatory lymphokines produced by B cells themselves. The possibility that sIg cross-linking is the sole stimulus involved in the induction of B cell proliferation seems unlikely, because B cells cultured at low density are stimulated poorly, if at all, to synthesize DNA by even high concentrations of anti-Ig antibodies.[55] The use of F(ab')₂ fragments of anti-Ig antibodies or particle-bound anti-Ig antibodies instead of intact soluble anti-Ig antibodies lowers the concentration of cells required for the induction of DNA synthesis.[143] Even under such circumstances ³H-thymidine incorporation per cultured cell declines as input cell number is decreased. ³H-thymidine incorporation by "purified" B cells cultured at low cell density with anti-Ig antibodies is greatly enhanced by addition of the monokine IL1 and the T lymphokine BCGF-I (recently redesignated BSF-pl) to the culture medium.[55,56] While this observation is compatible with the view that contaminating non-B cells in high density cultures of "purified" B cells produce factors required for the induction of DNA synthesis, the alternative view that these cytokines act synergistically with anti-Ig antibodies to stimulate B cell activation but do not play a role in anti-Ig-induced B cell activation in high cell density cultures is equally feasible.

When single B cells are cultured with anti-Ig antibodies, they fail to generate cell clones, even in the presence of helper cytokines, although some "T-independent" antigens or a mixture of LPS and dextran sulfate can induce clone formation.[57-59] Rather than indicate that B cells require signals other than sIg cross-linking and lymphokines to be stimulated to synthesize DNA, this result may indicate that B cells require additional signals to undergo repeated cycles of cell division, since the only indication of a positive response in this system, clone formation, requires the generation of considerable numbers of cells.

The possibility that high B cell concentrations are required for anti-Ig antibody induction of B cell DNA synthesis, because B cells themselves produce lymphokines that synergize with sIg cross-linking signals to promote activation, is suggested by the observation that the cloned B cell lymphoma, NBL, will proliferate in response to anti-μ antibody only when cultured at relatively high cell density unless additional stimuli are added to the culture medium.[60] This observation is consistent with reports that several B cell tumors, including NBL, have been shown to produce molecules that have activating or growth-stimulating effects on B cells.[60-62] The concept that anti-Ig antibody stimulated B cells may produce their own activation factor(s) is also consistent with the observation that GaMδ antibody is equally effective at stimulating B cell DNA synthesis in nude and conventional mice.[28] Recent observations also suggest that direct B cell-to-B cell interactions, promoted by anti-Ig antibodies, play a role in sIg cross-linking-dependent B cell activation, since incorporation of an agent in the culture medium that limits cell-to-cell contact inhibits the induction of DNA synthesis.[63] It seems unlikely, however, that anti-Ig antibodies activate B cells solely by promoting B cell-to-B cell contact, as many other antibodies that bind to B cell surface markers are incapable of stimulating B cell activation (see below).

3. Comparison of Abilities of Anti-Ig Antibodies and T-Dependent Antigens to Activate Antigen-Specific B Cells

If B cells can be activated by cross-linking of their sIg, it would be anticipated that purified populations of hapten-specific B cells would be similarly activated by a hapten-T-dependent antigen conjugate capable of cross-linking B cell sIg. Experiments from two laboratories, however, suggest that such conjugates enhance B cell sIa expression less than anti-Ig antibodies and, unlike anti-Ig antibodies, are incapable of stimulating B cell DNA synthesis, even in the presence of BCGF-I and IL1.[64-66] There are several possible explanations for this discrepancy. First, induction of proliferation might require the binding of a ligand to a specific site on the sIg molecule distinct from its antigen binding site. This seems unlikely, since anti-Ig antibodies specific for light chain, both murine δ chain domains, the four different μ chain domains, and even idiotypic determinants have all been found to stimulate B cell proliferation.[24-26,28,29,37,46,67,140] Secondly, the immunoglobulin nature of anti-Ig antibodies might in some way contribute to the induction of B cell mitogenesis. This also seems unlikely since hapten-goat IgG conjugates are no more effective than other hapten-T-dependent antigen conjugates at stimulating DNA synthesis by hapten-specific B cells.[144] A third possibility is that ligand-sIg interactions must be high avidity interactions if the degree of sIg cross-linking required for the induction of B cell proliferation is to occur. This is consistent with the finding that the monoclonal anti-δ antibody 10-4.22, which binds IgD with relatively low avidity and fails to precipitate serum IgD in an Ouchterlony plate, is much less effective at stimulating B cell DNA synthesis than is Hδᵃ/1, a more avid, precipitating anti-δ antibody.[37,53] This is despite the fact that both 10-4.22 and Hδᵃ/1 bind to the same allo-determinant on the Fc domain of the IgD molecule.[53] Two additional allospecific anti-δ monoclonal antibodies, AMS 15 and AF4, which fail to precipitate IgD in Ouchterlony, also fail to induce B cells to synthesize DNA, while two other monoclonal anti-δ antibodies, AMS 28 and AMS 9.1, which precipitate IgD as well as Hδᵃ/1, also stimulate B cell DNA synthesis.[68,145] No report has been published, however, that compares the ability of hapten-T-dependent antigen conjugates to activate B cells that bind these conjugates with high or low avidity. Even if the interpretation offered here is correct, it appears that most antigen-specific B cells are not directly activated to proliferate in vitro by T-dependent antigens. A direct sIg-ligand interaction could thus act as a mechanism to select those B cells that most avidly bind antigen. Alternatively, antigen-induced cross-linking of B cell sIg could act together with other stimulatory signals to induce B cell proliferation[65,69,146] or could have little role in B cell activation other than the focusing of polyclonal activator substances or antigen-specific T cell help onto the B cell.[8] Finally, since in vitro experiments with antigen-specific B cells have been performed at relatively low cell densities, while B cells are normally tightly packed together in vivo and B cells are most easily stimulated to proliferate by sIg cross-linking stimuli when cultured at high cell density, it is possible that antigen-induced B cell sIg cross-linking can induce B cell activation in vivo, even though in vitro B cell activation by the same stimulus is very limited.

IV. INTERACTIONS BETWEEN B CELL sIg AND THE IgGFc RECEPTOR

As noted above, one difference between the ways in which antigen and anti-Ig antibodies interact with antigen-specific B lymphocytes is that the latter often interact strongly with the B cell IgGFc receptor. Such interactions, however, generally have an inhibitory, rather than a stimulatory, effect on B cell activation. This inhibitory effect has been shown to block anti-Ig antibody stimulation of B cell DNA synthesis in a number of in vitro studies. Such studies have demonstrated that (1) F(ab')₂ fragments of rabbit anti-Ig antibodies are better able to stimulate B cell proliferation than are intact rabbit antimouse Ig antibody molecules;[70-72] (2) the monoclonal anti-IgGFc receptor antibody, 24G2,[73] which binds tightly to the IgGFc receptor without modulating it and blocks its ability to bind rabbit IgG,[74] enhances the mitogenic effects of intact rabbit anti-Ig antibodies, but not F(ab')₂ fragments of these antibodies;[75] and (3) intact goat antimouse Ig antibodies, which bind less well to the B cell IgGFc receptor than intact rabbit antimouse Ig antibodies,[76] are usually more mitogenic for B cells than are the rabbit antibodies and have their mitogenicity enhanced less by 24G2 than do the rabbit antibodies.[143] The IgGFc receptor-mediated inhibitory effects of rabbit anti-Ig antibodies are not noted when these antibodies are linked to a particle such as agarose beads.[23,72] One interpretation of these observations is that extensive cross-linking of the IgGFc receptor, which occurs when soluble anti-Ig antibodies cross-link sIg with the IgGFc receptor and cap both, but does not result from interactions between B cells and agarose-bound anti-Ig antibody or from interactions between B cells and 24G2[140] is required to inhibit the B cell activating effects of anti-Ig antibodies. Interestingly, IgGFc receptor inhibition of anti-Ig antibody-induced B cell activation is much more difficult to demonstrate in vivo than in vitro, since rabbit and goat anti-IgD antibodies have similar in vivo B cell activating effects and these effects are not enhanced by 24G2.[140] This correlates with the observation that the B cell IgGFc receptor is much more extensively modulated (as detected by immunofluorescence staining with 24G2) in vitro by rabbit antimouse IgD antibody than it is in vivo.[140] It is uncertain whether this difference reflects in vivo blocking by autologous plasma IgG of an interaction between rabbit antimouse IgD antibody and the B cell IgGFc receptor, or some other difference between the in vivo and in vitro environments.

V. COMPARISON OF THE STIMULATORY EFFECTS OF ANTI-Ig ANTIBODIES AND OTHER ANTI-B CELL ANTIBODIES

The observation that the cross-linking of B cell sIg by many different polyclonal or monoclonal anti-Ig antibodies suggested the possibility that similar interactions between non-Ig B cell surface molecules and antibodies to these molecules would have a similar effect. However, monoclonal antibodies to the IgGFc receptor (24G2),[73] class I and class II histocompatibility antigens, ThB (17C9),[77] and a number of other less well-characterized B cell markers have, in our hands, failed to induce B cells to increase their size, expression of sIa, or rate of DNA synthesis.[147] The one exception to this is a monoclonal allo-antibody to Lyb2,[78] which differs from anti-Ig antibodies in that its monovalent Fab fragment works as well as the intact molecule at inducing B cell pro-liferation in vitro.[79] In addition, unlike anti-δ antibodies, which are as effective in vivo as they are in vitro at stimulating B cell DNA synthesis,[28] anti-Lyb2 is ineffective at stimulating B cells to increase their size, expression of sIa, or rate of DNA synthesis in vivo.[148] This is despite (or perhaps, because of) the fact that anti-Lyb2 is actually more effective at modulating Lyb2 from the B cell surface in vivo than in vitro.[140] Thus, it can be concluded that the cross-linking of B cell sIg has special effects on B cell acti-vation that are not reproduced by the cross-linking of other B cell surface molecules.

VI. IN VIVO-IN VITRO DIFFERENCES IN B CELL ACTIVATION BY ANTI-Ig ANTIBODIES

For the most part in vivo studies of B cell activation in mature mice have been performed with anti-δ antibodies, since anti-μ or antilight-chain antibodies form com-plexes with the large quantities of immunoglobulin that are present in plasma. Goat, rabbit, and monoclonal-allospecific mouse antimouse δ antibodies each stimulate sim-ilar increases in the size, Ia expression, and rate of DNA synthesis by B cells of normal, mature mice in vivo and in vitro. Somewhat different effects have been seen, however, in the in vitro and in vivo response to GaMδ of B cells from mice that have the *xid* immune defect characteristic of CBA/N mice. *Xid* B cells stimulated in vitro with GaMδ antibody increase considerably in their expression of sIa and demonstrate size increases similar to those seen by normal B cells, but unlike normal B cells, fail to synthesize DNA.[29] In vivo, the increases in sIa expression and size of B cells from GaMδ-injected *xid* mice are considerably below those seen in GaMδ-injected normal mice, yet, in contrast to what is seen in vitro, some *xid* B cells are stimulated in vivo by GaMδ to synthesize DNA.[22] Thus, GaMδ is considerably less effective in vivo at inducing resting *xid* B cells to enter G_1 phase of the cell cycle than it is in vitro, but is more effective at inducing a subpopulation of *xid* B cells to synthesize DNA.

An even more marked difference between the in vitro and in vivo B cell activating effects of anti-Ig antibodies concerns the ability of GaMδ antibody to induce a B cell receptor for the lymphokine interleukin 2 (IL2). Incubation of purified B cells in vitro with soluble GaMδ antibody or GaMδ-agarose beads for as little as 16 hr induces considerable IL2 receptor expression by most of these cells,[139] as detected by staining with a monoclonal anti-IL2 receptor antibody.[80] In contrast, GaMδ fails to induce detectable B cell IL2 receptor expression in vivo. This difference does not result from in vivo saturation or modulation of the B cell IL2 receptor by endogenous IL2, since (1) some T cells from GaMδ-injected mice express easily detectable IL2 receptor; (2) addition of IL2 to in vitro cultures of GaMδ-stimulated B cells does not inhibit their expression of the IL2 receptor; and (3) similar in vivo results are seen in normal and in nude mice.[149] The in vitro-in vivo differences in IL2 receptor induction also do not result from in vivo-in vitro differences in GaMδ concentration, since GaMδ concentra-tions as low as 1 μg/mℓ stimulate B cells to express the IL2 receptor in vitro, while injection of mice with 1 mg of GaMδ fails to stimulate B cell IL2 receptor expression. This difference is more than likely the result of either a suppressive effect that is present in vivo but not in vitro, or of a stimulatory effect induced by culturing B cells in vitro. Mouse serum constituents may be partially responsible for the in vitro-in vivo differ-ence, since inclusion of 10% serum or plasma from normal or nude mice in cultures of B cells with GaMδ inhibits IL2 receptor expression (but not increased Ia expression) by these cells. Since IL2 has been shown to promote growth and differentiation of B cells in vitro,[81-84] the in vitro-in vivo difference in the effects of GaMδ on B cell IL2 receptor expression suggests that the stimuli required for the generation of clones of Ig-secreting cells in vivo may differ from, or be in addition to, those required in vitro.

VII. ROLES OF B CELL sIg IN THE INDUCTION OF ANTIBODY PRODUCTION

While anti-Ig antibodies are effective at activating B lymphocytes and stimulating these cells to synthesize DNA, these antibodies fail in the absence of additional stimuli to induce B cells to secrete Ig.[85] There is little doubt, however, that anti-Ig antibodies enhance the capacity of other signals to induce B cells to differentiate into antibody-secreting cells. B lymphocytes cultured in vitro with some supernatants of mitogen-stimulated T cells generate many more Ig-secreting cells than do B cells cultured with the same T cell culture supernatants in the absence of anti-Ig antibodies.[85,86] Anti-Ig antibodies that have been bound to a solid substrate and F(ab')$_2$ fragments of soluble anti-Ig antibodies are considerably more effective at stimulating antibody production than are intact, soluble anti-Ig antibodies, presumably because extensive cross-linking of the B cell IgGFc receptor inhibits the differentiation of B cells into Ig-secreting cells.[75,85,86] The ability of anti-Ig antibodies to synergize with activated T cell culture supernatants in the induction of Ig secretion probably relates, in part, to the induction by anti-Ig antibodies of B cell receptors for two lymphokines, Con A TRF and IL2, that are important for the induction of Ig secretion in this system.[41,81,83,84,86,87] Anti-Ig antibodies also have a stimulatory effect on the induction of Ig secretion in in vitro systems in which B cells are stimulated with relatively purified lymphokines. Low density B cell cultures, initially stimulated to proliferate with anti-μ antibody, IL1, and BCGF-1, are induced to secrete Ig by the addition of two B cell differentiation factors, one of which may contain or be identical to IL2.[88,150] Anti-Ig antibodies contribute substantially to the generation of Ig-secreting cells in systems in which helper lympho-kines plus lipopolysaccharide or helper T cells specific for B cell antigens are used to induce the polyclonal generation of Ig-secreting cells.[89,151] In some of these systems LPS and anti-Ig antibodies have been shown to act synergistically in the induction of the expression of high affinity B cell receptors for IL2.[83] In other in vitro studies, however, in which B cells were cultured with helper T cell lines specific for B cell antigens, anti-Ig antibodies or other ligands capable of cross-linking B cell sIg were not found to contribute to the generation of Ig-secreting cells.[90] Reasons for this discrepancy are uncertain. Since stimulatory signals transmitted through sIg contribute more to the generation of Ig-secreting cells from populations of small, resting B cells than from populations of larger, more activated cells, signals transmitted by the cross-linking sIg may, in these systems, have their primary role in the initial activation of B cells.[89,91] Thus, the importance of such signals may be evident only when populations of small B cells are studied. In addition, since different helper T cell lines have been used in different studies, it is possible that some T cell lines provide help potent enough to make additional stimulatory signals transmitted through sIg superfluous, while other T cell lines that provide less potent help require the additional B cell stimulation produced by sIg cross-linking to induce Ig secretion. A third possibility is that the requirements for the induction of Ig secretion in some in vitro systems are less stringent than the requirements in other in vitro or in vivo systems. The results of studies of B cell IL2 receptor induction and the activating effects of anti-Lyb2 antibody, noted above, are both consistent with the view that the requirements for induction of B cell proliferation and differentiation may be different in vivo than in vitro.

These in vivo-in vitro differences in B cell activation requirements make it worthwhile to try to investigate the role of B cell sIg in the induction of Ig secretion in vivo, despite the obvious difficulties in vivo in controlling stimuli and cell populations. While anti-Ig antibodies fail to stimulate spleen cells to secrete Ig in vitro, and even in some circumstances, inhibit antigen- or mitogen-stimulated Ig secretion,[72,92-95] mice injected with affinity purified heterologous anti-δ antibodies, or some monoclonal mouse allo-anti-δ antibodies, demonstrate large polyclonal IgG1 antibody responses 6 to 9 days after anti-δ antibody injection.[96] In BALB/c mice injected with optimal doses of GaMδ antibody, a three- to eightfold increase in splenic B cell number is seen 7 to 8 days after injection, and up to 50% of these cells secrete IgG1.[28,45,47,96,97] This results in a 50- to 100-fold increase in serum IgG1 levels.[47] Unlike the in vivo stimulation of B cell proliferation and receptor expression by GaMδ antibodies, however, the induction of Ig secretion by GaMδ antibodies is T dependent, since it does not occur in nude mice.[96] Furthermore, in vivo anti-δ antibody induction of polyclonal IgG1 secretion is dependent upon the presence of T lymphocytes capable of recognizing the injected anti-δ antibody as a foreign antigen. In other words, GaMδ antibody fails to induce polyclonal Ig secretion in mice previously exposed to goat IgG, and monoclonal mouse IgG2a anti-δ antibodies induce polyclonal IgG secretion only when the anti-δ antibodies themselves bear IgG2a allotypic determinants different from those characteristic of the mice into which they are injected.[96,145] A role for T cells in the induction of polyclonal Ig secretion in this system is also suggested by the observation that increased splenic T cell DNA synthesis is observed as early as 2 days after GaMδ injection and reaches plateau levels by day 3.[45] It is not known, however, what percent of the proliferating T cells are specific for goat IgG determinants. The crucial role of B cell sIgD in the induction of polyclonal Ig secretion in this system is indicated by the inability of control, heterologous antibodies, monoclonal mouse antibodies that fail to bind to B cells, or rat monoclonal antibodies that bind to non-Ig B cell determinants to induce polyclonal Ig secretion in vivo.[53,96,97,140] Taken together with the observations about T-independent B cell activation by anti-Ig antibodies discussed earlier, these in vivo studies performed with GaMδ suggest that (1) the cross-linking of sIgD by anti-δ antibodies polyclonally activates B cells to increase their expression of receptors important for their interaction with both T cells and stimulatory lymphokines secreted by T cells and to synthesize DNA; (2) these activated B cells present antigenic determinants on the GaMδ molecules to helper T cells specific for these determinants and activate these T cells;[98-100] and (3) these activated goat IgG-specific T cells, lymphokines secreted by these T cells, and/or other cells activated by these T cells stimulate B cells to secrete IgG1.[53,96] It is not known what role, if any, macrophages, dendritic cells, or other antigen-presenting cells have in these processes. We also do not know whether the contribution of GaMδ-induced B cell activation in the induction of polyclonal Ig secretion is restricted to antigen presentation to and activation of helper T cells, or whether the GaMδ-activated B cells also are more responsive than resting B cells to the activated T cells and their secreted products. In addition, the observations described above do not clarify whether the T help responsible for polyclonal Ig secretion in the in vivo GaMδ system is antigen- (goat IgG) specific help that is focused onto B cells by GaMδ bound to their sIgD, nonantigen-specific help, or both. To differentiate between the role played by antigen-specific and nonspecific T help, experiments were performed in which GaMδ antibody was neutralized by large doses of IgD at various times after injection to limit the period of time during which GaMδ could focus goat IgG-specific T help onto B cells. Results of these studies indicated that (1) neutralization of GaMδ 3 days after injection has a strong inhibitory effect on the development of a polyclonal IgG1 antibody response; (2) this inhibition can be overcome by injecting mice with normal goat IgG at the time GaMδ is neutralized; and (3) neutralization of GaMδ earlier than 3 days after injection inhibits the development of a polyclonal antibody response even in mice injected with large doses of normal goat IgG.[45] Since GaMδ (but not normal goat IgG) could focus goat IgG-specific T help polyclonally onto B cells, while both GaMδ and normal goat IgG could induce activated goat IgG-specific T cells to generate helper factors, these data indicate that antigen-nonspecific T help has an important role in the stimulation of polyclonal antibody production in this system and suggest that either antigen-specific (focused) T help has no role in the generation of polyclonal antibody production in this system or the role of such help is restricted to the first 3 days after GaMδ injection. Although we cannot exclude either possibility, we favor the former since B cell activation appears to proceed similarly in nude and normal mice for the first 3 days after GaMδ injection.[96]

VIII. DIFFERENCES IN THE ROLES OF B CELL MEMBRANE IgM AND IgD IN B CELL ACTIVATION

One of the more frustrating areas of research in B cell immunology has been the attempt to find a biological purpose for the possession of both sIgM and sIgD by most mature, virgin, resting B cells. Since the variable regions of both membrane isotypes on a single B cell clone are encoded by the same DNA sequence,[101] and thus have identical antigen binding sites,[6,16] B cells with either isotype should be as capable of binding an antigen as B cells that bear both isotypes. Furthermore, the observation that μ and δ chains have identical intracytoplasmic carboxy-terminal segments[102] suggests that both isotypes will interact with identical cytoplasmic molecules and may thus transmit identical messages to B lymphocytes. Studies of the cellular and molecular biology of IgM and IgD have, however, yielded information about these molecules that may be regarded as clues to their functions, and which provide a framework for interpreting comparative studies performed with anti-μ and anti-δ antibodies. These "clues" include observations that (1) serum IgD levels are much lower than IgM levels;[103,104] (2) B cells can express both sIgM and sIgD without having to excise the gene for either heavy chain C region;[105] (3) B cells acquire sIgM earlier in ontogeny than sIgD;[15,106] (4) most mature, resting, virgin, murine B cells bear as much as tenfold more sIgD than sIgM;[14,15,107] (5) once activated to synthesize DNA, B cells lose most or all of their sIgD, but retain sIgM;[108-110] (6) the synthetic rate of sIgM is considerably greater than that of sIgD, even when small B cells from mature mice are analyzed;[111] (7) the half-life of sIgD on mature, resting B cells is considerably greater than that of sIgM;[111] (8) IgD has an extensive hinge region[112,113] that should allow its antigen binding sites considerable independent movement[114] (and which makes it susceptible to enzymatic cleavage),[115] while sIgM, which lacks a hinge region[116] and is relatively stable to enzymatic cleavage, should have little segmental flexibility; (9) sIgM⁺IgD⁻ B cells do not recirculate (at least in the rat), while the sIgM⁺IgD⁺ B cell population contains the great majority of recirculating B lymphocytes;[117] (10) sIgM⁺IgD⁻ B cells are resistant to in vitro activation by either anti-δ or anti-μ antibodies, while both anti-δ and anti-μ antibodies have their greatest mitogenic effects on B cells that have a sIgM dull, sIgD bright phenotype;[25,46,118,119] and (11) mice suppressed from birth with anti-δ antibodies (which have few or no sIgD⁺ B cells and increased numbers of sIgM bright B cells)[120,121] make relatively normal primary and secondary IgM and IgG responses to both high and low epitope density forms of T-dependent and relatively T-independent antigens,[118,122] however, they appear in preliminary studies to have increased susceptibility to at least some infectious agents.[123,124] In addition to these generally agreed-upon observations, some, but not all, investigators have reported that (1) soluble anti-μ antibodies are more effective than soluble anti-δ antibodies at inducing some B cell populations to enter the cell cycle and synthesize DNA;[39,75] (2) antigen-stimulated sIgM⁺IgD⁻ B cells fail to generate in vitro antibody responses;[125,126] (3) removal of sIgD from sIgM⁺IgD⁺ B cells makes them more susceptible to tolerance induction;[127-129] and (4) anti-δ antibodies block in vitro primary antibody responses to T-dependent and type 2 antigens but not type 1 antigens, while anti-μ antibodies block all in vitro antibody responses.[93,94,130] These observations have stimulated the development of a number of theories about the relative functions of sIgM and sIgD. It has been proposed that (1) sIgD is required for responses to low epitope density antigens;[114] (2) sIgD is required for responses to T-dependent antigens and TI-2 antigens, while sIgM is required for responses to all antigens;[93,130] (3) sIgD has a special role in idiotype-mediated immune regulation;[131] and (4) sIgD confers B cells with the ability to resist tolerance induction.[127-129] None of these theories has been proven, and some seem inconsistent with the results of studies performed with mice suppressed from birth with anti-δ antibody.[118,120,122] The observations cited above, however, as well as observations made with the in vivo GaMδ system, recently led to the proposal that there is a division of labor between sIgM and sIgD in antigen-specific B cell activation, such that IgD is the predominant sIg isotype involved in direct, T-independent, sIg cross-linking-dependent B cell activation, while IgM is the sIg isotype most involved in the focusing of antigen-specific T help onto B cells.[53] Such a division of labor was suggested by the belief that sIgD, because of its extended hinge region and presumed high segmental flexibility, is more likely to be functionally divalent and, thus, more likely to be cross-linked by antigen, than is hinge region-deficient sIgM.[114] In such a model, sIgD would function predominantly in early B cell activation. The loss of sIgD by activated B cells would limit extensive antigen-induced sIg cross-linking of already activated B cells, which has been found in some in vivo and in vitro systems to inhibit the differentiation of B cells into antibody-secreting cells.[53,93-95,114,130] The retention of sIgM by activated B cells would provide a mechanism for focusing antigen-specific T cell help or antigen-associated B cell polyclonal activator substances onto these cells. This model is supported by studies with GaMδ-injected mice which demonstrate that (1) although B cells are stimulated by GaMδ-induced sIgD cross-linking to synthesize DNA and by GaMδ-induced T help in this system to polyclonally differentiate into IgG1 secreting cells, only those B cells specific for goat IgG undergo a large (several hundred-fold) increase in clone size; (2) the greater clonal expansion of goat IgG-specific B cells derives from the ability of these cells, but not B cells polyclonally, to bind GaMδ (or any goat IgG) with their sIgM; (3) extensive clonal expansion of hapten-specific B cells in GaMδ-injected mice requires the focusing of goat IgG-specific T help onto hapten-specific B cells via a hapten-goat IgG conjugate bound to their sIgM rather than simply binding hapten to sIgM or cross-linking sIgM by hapten-carrier conjugates. That is, mice injected with GaMδ plus hapten conjugated to goat IgG make a much larger antihapten antibody response than mice injected with GaMδ plus hapten bound to carriers other than goat IgG; (4) maximal B cell clonal expansion requires that antigen-specific T help be focused onto B cells for at least a 5-day period.[32] The early loss of sIgD by activated B cells requires that this prolonged focusing of T help onto B cells be mediated through sIgM.

IX. PRESENT KNOWLEDGE, PRESENT IGNORANCE, AND FUTURE QUESTIONS

Experiments performed in vitro and in vivo with anti-Ig antibodies have demonstrated that sIg can be involved in direct B cell activation, in the presentation of antigen to and the activation of antigen-specific T lymphocytes, and in the focusing of antigen-specific help onto B cells. Studies performed with cytokines, bacterial products, and organic compounds have similarly identified a large number of mediators that have the potential, under defined experimental conditions, to activate B cells, and induce B cell proliferation, Ig secretion, and isotype switching.[8,32,33,35,43,55-57,67,69,81,82,86-88,132-138] Under some experimental conditions, signals mediated through sIg do not appear to be required for the induction of B cell proliferation, clonal expansion, or antibody secretion.[8,90,133,137] What is still unclear is the relevance of these different B cell activation systems to the generation of an antibody response to an antigenic challenge. Multiple questions remain unanswered: To what extent are the direct B cell activating effects of anti-Ig antibodies reproduced by the interaction of antigen with the sIg of antigen-specific B cells? To what extent is sIg cross-linking-mediated B cell activation dependent upon the high B cell density seen in vivo, but not in many in vitro systems? To what extent do sIg cross-linking-mediated activation signals act additively or synergistically with other activation signals? To what extent do sIg cross-linking-mediated activation signals induce B cell receptors or otherwise modify the B cell to allow it to respond to subsequent activating signals? Does sIg cross-linking-dependent B cell activation contribute to the generation of antibody-secreting cells in vivo only by stimulating the generation of T cell help, or also by making B cells more responsive to such help? Does the presence of two sIg isotypes on the same B cell increase the flexibility of the immune system? If so, how? With regard to the roles of cytokines and cell-mediated T help in B cell activation, it needs to be determined whether these agents achieve concentrations in vivo that have been demonstrated to have B cell activating effects in vitro, whether such concentrations are achieved at times during an immune response when these agents have their putative effects, and whether these agents have effects in the in vivo milieu that resemble those seen in vitro. With regard to the different pathways of B cell activation that have been detected in vitro, it remains to be determined which of these pathways can function in vivo. Should multiple in vivo B cell activation pathways be found to exist, as seems likely, it will be necessary to determine when these multiple pathways represent redundant systems that allow a host to survive infection with a pathogen when one pathway fails, when these pathways function separately to allow responses by different B cell subsets to different kinds of antigens,[21] and when these pathways function additively or synergistically to maximize the size and speed of an antibody response to a potentially lethal infectious organism.[53] It is clear that neither in vitro nor in vivo systems can be used alone to answer all of these questions. The challenge to generate in vitro systems that better reflect physiological conditions and in vivo systems that are sufficiently well defined to approach these problems persists.

ACKNOWLEDGMENTS

Many of the studies described in this review were supported by USUHS protocol numbers RO8308, RO8347, and CO7305; Office of Naval Research Contracts N000148AF00001 and M000950001.1030; National Science Foundation Grants No. PCM-8215682 and PCM-8217009; and National Institutes of Health Grant No. 1RO1AI21328-01. We are grateful for the expert technical assistance of Mrs. Joanne Smith, Ms. Nelly Villacreses, Mr. Chander Sarma, Ms. Mary Schaefer, and Mr. Neil Hardigan; for the secretarial assistance of Ms. Janet Thomson; and for the ideas and help of our many collaborators, especially Drs. Irwin Scher, Linda Muul, Richard Weber, Diana Goroff, Donna Sieckmann, and Craig Thompson. The opinions and assertions contained herein are those of the authors and are not to be construed as official or reflecting the views of the Department of Defense or the Uniformed Services University of the Health Sciences. Experiments performed in our laboratories and reported herein were conducted according to the principles set forth in the "Guide for the Care and Use of Laboratory Animals", Institute of Animal Resources, National Research Council, DHEW Publ. No. (NIH) 78-23.

REFERENCES

1. Sell, S. and Gell, P. G. H., Studies on rabbit lymphocytes *in vitro*. I. Stimulation of blast transformation with an anti-allotype serum, *J. Exp. Med.*, 122, 423, 1965.
2. Pernis, B., Forni, L., and Amante, L., Immunoglobulins as cell receptors, *Ann. N.Y. Acad. Sci. U.S.A.*, 190, 420, 1971.
3. Raff, M. C., Sternberg, J., and Taylor, R., Immunoglobulin determinants on the surface of mouse lymphoid cells, *Nature (London)*, 225, 553, 1970.
4. Salsano, F., Froland, S. S., Natvig, V. B., and Michaelson, T. E., Same idiotype of B lymphocyte membrane IgD and IgM. Evidence for monoclonality of chronic lymphocytic leukemia cells, *Scand. J. Immunol.*, 3, 841, 1974.
5. Fu, S. M., Winchester, R. J., and Kunkel, H. G., Similar idiotypic specificity for the membrane IgD and IgM of human B lymphocytes, *J. Immunol.*, 114, 250, 1975.
6. Pernis, B., Brouet, J. C., and Seligmann, M., IgD and IgM on the membrane of lymphoid cells in the macroglobulinemia. Evidence of identity of membrane IgD and IgM antibody activity in a case with anti-IgG receptors, *Eur. J. Immunol.*, 4, 776, 1974.
7. Bretscher, P. A. and Cohn, M., A theory of self-nonself discrimination, *Science*, 169, 1042, 1970.
8. Coutinho, A. and Möller, G., Immune activation of B cells: evidence for one non-specific triggering signal not delivered by the Ig receptors, *Scand. J. Immunol.*, 3, 133, 1974.
9. Van Boxel, J. A., Paul, W. E., Terry, W. D., and Green, I., IgD-bearing human lymphocytes, *J. Immunol.*, 109, 648, 1972.
10. Rowe, D. S., Hug, K., Forni, L., and Pernis, B., Immunoglobulin D as a lymphocyte receptor, *J. Exp. Med.*, 138, 965, 1973.
11. Knapp, W., Bolus, R. L. H., Radl, J., and Higmans, W., Independent movement of IgD and IgM molecules on the surface of individual lymphocytes, *J. Immunol.*, 111, 1295, 1973.
12. Fu, S. M., Winchester, R. J., and Kunkel, H. B., Occurrence of surface IgM, IgD and free light chains on human lymphocytes, *J. Exp. Med.*, 139, 451, 1974.
13. Leslie, G. A. and Armen, R. C., Structure and biological functions of human IgD. III. Phylogenetic studies of IgD, *Int. Arch. Allergy*, 46, 191, 1974.
14. Hardy, R. R., Hayakawa, K., Haaijman, J., and Herzenberg, L. A., B cell subpopulations identifiable by two-color fluorescence analysis using a dual-laser FACS, *Ann. N.Y. Acad. Sci.*, 399, 112, 1982.
15. Scher, I., Titus, J. A., and Finkelman, F. D., The ontogeny and distribution of B cells in normal and mutant immune defective CBA/N mice: two parameter analysis of surface IgM and IgD, *J. Immunol.*, 130, 619, 1983.
16. Goding, J. W. and Layton, J. E., Antigen-induced co-capping of IgM and IgD-like receptors on murine B cells, *J. Exp. Med.*, 144, 852, 1976.
17. Scher, I., Steinberg, A. D., Berning, A. K., and Paul, W. E., X-linked B lymphocyte defect in CBA/N mice. II. Studies of the mechanisms underlying the immune defect, *J. Exp. Med.*, 142, 637, 1975.
18. Weiner, H. L., Moorhead, J. W., and Claman, H. N., Anti-immunoglobulin stimulation of murine lymphocytes. I. Age dependency of the proliferative response, *J. Immunol.*, 116, 1656, 1976.
19. Bona, C., Mond, J. J., Stein, K. E., House, S., Lieberman, R., and Paul, W. E., Immune response to levan. III. The capacity to produce anti-inulin antibodies and cross-reactive idiotypes appears late in ontogeny, *J. Immunol.*, 123, 1484, 1979.
20. Greenstein, J. L., Lord, E., Kappler, J. W., and Marrack, P. C., Analysis of the response of B cells from CBA/N defective mice to nonspecific T cell help, *J. Exp. Med.*, 154, 1608, 1981.
21. Asano, Y., Singer, A., and Hodes, R. J., Role of the major histocompatibility complex in T cell activation of B cell subpopulations: major histocompatibility complex-restricted and -unrestricted B cell responses are mediated by distinct B cell subpopulations, *J. Exp. Med.*, 154, 1100, 1981.
22. Muul, L. M., Scher, I., Mond, J. J., and Finkelman, F. D., Polyclonal activation of the murine immune system by an antibody to IgD. IV. *In vivo* activation of B lymphocytes from immune defective CBA/N mice, *J. Immunol.*, 131, 2226, 1983.
23. Parker, D. C., Stimulation of mouse lymphocytes by insoluble anti-mouse immunoglobulin, *Nature (London)*, 258, 361, 1975.
24. Sieckmann, D. G., Asofsky, R., Mosier, D. E., Zitron, I. M., and Paul, W. E., Activation of mouse lymphocytes by anti-immunoglobulin. I. Parameters of the proliferative response, *J. Exp. Med.*, 147, 814, 1978.
25. Sieckmann, D. G., Scher, I., Asofsky, R., Mosier, D. E., and Paul, W. E., Activation of mouse lymphocytes by anti-immunoglobulin. II. A thymus-independent response by a mature subset of B lymphocytes, *J. Exp. Med.*, 148, 1628, 1978.
26. Sidman, C. L. and Unanue, E. R., Requirements for mitogenic stimulation of murine B cells by soluble anti-IgM antibodies, *J. Immunol.*, 122, 406, 1979.

27. Mond, J. J., Sehgal, E., Kung, J., and Finkelman, F. D., Increased expression of I-region-associated antigen (Ia) on B cells after cross-linking of surface immunoglobulin, *J. Immunol.*, 127, 881, 1981.

28. Finkelman, F. D., Scher, I., Mond, J. J., Kung, J. T., and Metcalf, E. S., Polyclonal activation of the murine immune system by an antibody to IgD. I. Increase in cell size and DNA synthesis, *J. Immunol.*, 129, 629, 1982.

29. DeFranco, A. L., Raveche, E. S., Asofsky, R., and Paul, W. E., Frequency of B lymphocytes responsive to anti-immunoglobulin, *J. Exp. Med.*, 155, 1523, 1982.

30. Rajewsky, R., Shirrmacher, V., Nase, S., and Jerne, N. K., The requirement of more than one antigenic determinant for immunogenicity, *J. Exp. Med.*, 129, 1131, 1969.

31. Mitchison, N. A., The carrier effect in the secondary response to hapten-protein conjugates. II. Cellular cooperation, *Eur. J. Immunol.*, 1, 18, 1971.

32. Andersson, J. and Mechers, F., T cell-dependent activation of resting B cells: requirements for both nonspecific unrestricted and antigen-specific Ia-restricted soluble factors, *Proc. Natl. Acad. Sci. U.S.A.*, 78, 2497, 1981.

33. Jaworski, M. A., Shiozawa, C., and Diener, E., Triggering of affinity-enriched B cells. Analysis of B cell stimulation by antigen-specific helper factor or lipopolysaccharide. I. Dissection into proliferative and differentiative signals, *J. Exp. Med.*, 155, 248, 1981.

34. Singer, A. and Hodes, R. J., Mechanisms of T cell-B cell interaction, in *Annual Review of Immunology*, Paul, W. E., Fathman, C. G., and Metzger, H., Eds., Annual Review, Palo Alto, Calif., 1983, 211.

35. Roehm, N. W., Marrack, P., and Kappler, J. W., Helper signals in the plaque-forming cell response to protein-bound haptens, *J. Exp. Med.*, 158, 317, 1983.

36. Snow, E. C., Vitetta, E. S., and Uhr, J. W., Activation of antigen-enriched B cells. I. Purification and response to thymus-independent antigens, *J. Immunol.*, 130, 607, 1983.

37. Zitron, I. M. and Clevinger, B. L., Regulation of murine B cells through surface immunoglobulin. I. Monoclonal anti-δ antibody that induces allotype-specific proliferation, *J. Exp. Med.*, 152, 1135, 1980.

38. Julius, M. H., Heusser, C. H., and Hartmann, K.-V., Induction of resting B cells to DNA synthesis by soluble monoclonal anti-immunoglobulin, *Eur. J. Immunol.*, 14, 753, 1984.

39. Cambier, J. C. and Monroe, J. G., B cell activation. V. Differential signaling of B cell membrane depolarization, increased I-A expression, G_0 to G_1 transition, and thymidine uptake by anti-IgM and anti-IgD antibodies, *J. Immunol.*, 133, 576, 1984.

40. Weber, R. J. and Finkelman, F. D., Increased expression of transferrin receptor on B cells after *in vivo* cross-linking of surface IgD, *Fed. Proc. Fed. Am. Soc. Exp. Biol.*, 43, 1424, 1984.

41. Yaffe, L. J. and Finkelman, F. D., Induction of a B-lymphocyte receptor for a T cell-replacing factor by the cross-linking of surface IgD, *Proc. Natl. Acad. Sci. U.S.A.*, 80, 293, 1983.

42. Gallatin, W. M., Weissman, I. L., and Butcher, E. C., A cell-surface molecule involved in organ-specific homing of lymphocytes, *Nature (London)*, 304, 40, 1983.

43. Noelle, R., Krammer, P. H., Ohara, J., Uhr, J., and Vitetta, E. S., Increased expression of Ia antigens on resting B cells: a new role for B cell growth factor, *Proc. Natl. Acad. Sci. U.S.A.*, 81, 6149, 1984.

44. Thompson, C. B., Schaefer, M. E., Finkelman, F. D., Farrar, J., and Mond, J. J., T cell-derived B cell growth factor (BCGF) induces stimulation of both resting and activated B cells, *J. Immunol.*, 134, 369, 1985.

45. Finkelman, F. D., Smith, J., Villacreses, N., and Metcalf, E. S., Polyclonal activation of the murine immune system by an antibody to IgD. VII. Demonstration of the role of nonantigen-specific T help in *in vivo* B cell activation, *J. Immunol.*, 133, 550, 1984.

46. Sieckmann, D. G., Finkelman, F. D., and Scher, I., IgD as a receptor in signaling the proliferation of mouse B lymphocytes, *Cell. Immunol.*, 85, 1, 1984.

47. Finkelman, F. D., Smith, J., Villacreses, N., and Metcalf, E. S., Polyclonal activation of the murine immune system by an antibody to IgD. VI. Influences of doses of goat anti-mouse δ and normal goat IgG on B lymphocyte proliferation and differentiation, *Eur. J. Immunol.*, in press.

48. Weiner, H. L., Moorhead, J. W., Yamaga, K., and Kubo, R. T., Anti-immunoglobulin stimulation of murine lymphocytes. II. Identification of cell surface target molecules and requirements for cross linkage, *J. Immunol.*, 117, 1527, 1976.

49. DeFranco, A. L., Kung, J. T., and Paul, W. E., Regulation of growth and proliferation in B cell subpopulations, *Immunol. Rev.*, 64, 161, 1982.

50. Pure, E. and Vitetta, E., Induction of murine B cell proliferation by insolubilized anti-immunoglobulins, *J. Immunol.*, 125, 1240, 1980.

51. Mond, J. J., Schaefer, M., Smith, J., and Finkelman, F. D., Lyb5⁻ cells can be induced to synthesize DNA by culture with insolubilized anti-immunoglobulin but not with soluble anti-immunoglobulin, *J. Immunol.*, 131, 3107, 1983.

52. Ramanadham, M., Gollapud, S. V. S., and Kern, M., Fate of surface immunoglobulin during induction of lymphocyte proliferation, *J. Cell. Biochem.*, 24, 187, 1984.

53. Finkelman, F. D. and Vitetta, E. S., Role of surface immunoglobulin in B lymphocyte activation, *Fed. Proc. Fed. Am. Soc. Exp. Biol.*, 43, 2624, 1984.

54. Gollapudi, S. V. S., Ramanadham, M., and Kern, M., Soluble and insoluble anti-immunoglobulins of identical specificity activate different subpopulations of B lymphocytes, submitted for publication.

55. Howard, M., Farrar, J., Hilfiker, M., Johnson, B., Takatsu, K., Hamaoka, T., and Paul, W. E., Identification of a T cell-derived B cell growth factor distinct from interleukin 2, *J. Exp. Med.*, 155, 914, 1982.

56. Howard, M., Mizel, S. B., Lachman, L., Ansel, J., Honson, B., and Paul, W. E., Role of interleukin 1 in anti-immunoglobulin-induced B cell proliferation, *J. Exp. Med.*, 157, 1529, 1983.

57. Wetzel, G. D. and Kettman, J. R., Activation of murine B lymphocytes. III. Stimulation of B lymphocyte clonal growth with lipopolysaccharide and dextran sulfate, *J. Immunol.*, 126, 723, 1981.

58. Pike, B. L. and Nossal, G. J. V., A reappraisal of "T-independent" antigens. I. Effect of lymphokines on the response of single adult hapten-specific B lymphocytes, *J. Immunol.*, 132, 1687, 1984.

59. Nossal, G. J. V. and Pike, B. L., A reappraisal of "T-independent" antigens. II. Studies on single hapten-specific B cells from neonatal CBA/H or CBA/N mice fail to support classification into TI-1 and TI-2 categories, *J. Immunol.*, 132, 1696, 1984.

60. Ling, M., Livnat, D., Billai, P. S., and Scott, D. W., Lymphoma models for B-cell activation and tolerance. I. Conditions for the anti-μ dependent stimulation of growth in NBL, a nude B-cell lymphoma, *J. Immunol.*, 133, 1449, 1985.

61. Ambrus, J. and Fauci, A. S., Production of a B cell growth factor by human B cell lymphoma lines, *Fed. Proc. Fed. Am. Soc. Exp. Biol.*, 43, 1592, 1984.

62. Brooks, K. H., Uhr, J. W., and Vitetta, E. S., A B cell growth factor-like activity is secreted by cloned, neoplastic B cells, *J. Immunol.*, 133, 3133, 1984.

63. Spieker-Polet, H., Hagen, K., and Teodorescu, M., Requirements for intercellular contact in the activation of B lymphocytes by anti-immunoglobulin antibodies, *Fed. Proc. Fed. Am. Soc. Exp. Biol.*, 43, 1592, 1984.

64. Cambier, J. C., Monroe, J. G., and Neale, M. J., Definition of conditions that enable antigen-specific activation of the majority of isolated trinitrophenyl-binding B cells, *J. Exp. Med.*, 156, 1635, 1982.

65. Snow, E. C., Noelle, R. J., Uhr, J. W., and Vitetta, E. S., Activation of antigen-enriched B cells. II. Role of linked recognition in B cell proliferation to thymus-dependent antigens, *J. Immunol.*, 130, 614, 1983.

66. Noelle, R. J., Snow, E. C., Uhr, J. W., and Vitetta, E., Activation of antigen-specific B cells: the role of T cells, cytokines and antigen in the induction of growth and differentiation, *Proc. Natl. Acad. Sci. U.S.A.*, 80, 6628, 1983.

67. Melchers, F., Corbel, C., and Leptin, M., Requirements for B-cell stimulation, in *Progress in Immunology V*, Yamamura, Y. and Tada, T., Eds., Academic Press, Tokyo, 1983, 669.

68. Stall, A. M. and Loken, M. R., Allotypic specificities of murine IgD and IgM recognized by monoclonal antibodies, *J. Immunol.*, 133, 787, 1984.

69. Zubler, R. H. and Glasebrook, A. L., Requirement for three signals in T-independent (lipopolysaccharide-induced) as well as in T-dependent B cell responses, *J. Exp. Med.*, 155, 666, 1982.

70. Sidman, C. L. and Unanue, E. R., Control of B lymphocyte function. I. Inactivation of mitogenesis by interactions with surface immunoglobulin and Fc receptor molecules, *J. Exp. Med.*, 144, 882, 1976.

71. Scribner, D. J., Weiner, H. L., and Moorhead, J. W., Anti-immunoglobulin stimulation of murine lymphocytes. V. Age related decline in Fc receptor-mediated immunoregulation, *J. Immunol.*, 121, 377, 1978.

72. Vitetta, E. S., Pure, E., Isakson, P. C., Buck, L., and Uhr, J. W., The activation of murine B cells: the role of surface immunoglobulins, *Immunol. Rev.*, 53, 211, 1980.

73. Unkeless, J. C., Characterization of a monoclonal antibody directed against mouse macrophage and lymphocyte Fc receptors, *J. Exp. Med.*, 150, 580, 1979.

74. Katona, I. M., Urban, J. F., Jr., Titus, J. A., Stephany, D. A., Segal, D. M., and Finkelman, F. D., Characterization of murine lymphocyte IgE receptors by flow microfluirometry, *J. Immunol.*, 133, 1521.

75. Phillips, N. E. and Parker, D. C., Crosslinking of B lymphocyte Fcγ receptors and membrane immunoglobulin inhibits anti-immunoglobulin induced blastogenesis, *J. Immunol.*, 132, 627, 1984.

76. Tsay, D. D., Ogden, D., and Schlamowitz, M., Binding of homologous and heterologous IgG to Fc receptors on the fetal rabbit yolk sac membrane, *J. Immunol.*, 124, 1562, 1980.

77. Kung, J. T., Sharrow, S. O., Thomas, C. A., and Paul, W. E., Analysis of B lymphocyte differentiation antigens by flow microfluirometry, *Immunol. Rev.*, 69, 51, 1983.

78. Subbarao, B. and Mosier, D. E., Induction of B lymphocyte proliferation by monoclonal anti-Lyb2 antibody, *J. Immunol.*, 130, 2033, 1983.

79. Subbarao, B. and Mosier, D. E., Activation of B lymphocytes by monovalent anti-Lyb2 antibodies, *J. Exp. Med.*, 159, 1796, 1984.

80. Malek, T. R., Robb, R. J., and Shevach, E. M., Identification and initial characterization of a rat monoclonal antibody reactive with the murine interleukin 2 receptor-ligand complex, *Proc. Natl. Acad. Sci. U.S.A.*, 80, 5694, 1983.

81. Leibson, H. J., Marrack, P., and Kappler, J. W., B cell helper factors. I. Requirement for both interleukin 2 and another 40,000 mol. wt. factor, *J. Exp. Med.*, 154, 1681, 1981.

82. Mond, J. J., Thompson, C., Schaefer, M., Finkelman, F. D., and Robb, R., Immunoaffinity purified interleukin 2 (IL2) induces B cell growth, *Curr. Top. Microbiol. Immunol.*, 113, 102, 1984.

83. Zubler, R. H., Lowenthal, J. W., Erard, F., Hashimoto, N., Devos, R., and MacDonald, H. R., Activated B cells express receptors for, and proliferate in response to pure interleukin 2, *J. Exp. Med.*, 160, 1170, 1984.

84. Parker, D. C., Separable helper factors support B cell proliferation and maturation to Ig secretion, *J. Immunol.*, 129, 469, 1982.

85. Parker, D. C., Fothergill, J. J., and Wadsworth, D. C., B lymphocyte activation by insoluble anti-immunoglobulin: induction of immunoglobulin by a T cell-dependent soluble factor, *J. Immunol.*, 123, 931, 1979.

86. Pure, E., Isakson, P. C., Takatsu, K., Hamaoka, T., Swain, S. L., Dutton, R. W., Dennert, G., Uhr, J. W., and Vitetta, E. S., Induction of B cell differentiation by T cell factors. I. Stimulation of IgM secretion by products of a T cell hybridoma and a T cell line, *J. Immunol.*, 127, 1953, 1981.

87. Kishimoto, T., Miyake, T., Nishizawa, Y., Watanage, T., and Yamamura, Y., Triggering mechanism of B lymphocytes. I. Effect of anti-immunoglobulin and enhancing soluble factor on differentiation and proliferation of B cells, *J. Immunol.*, 115, 1179, 1975.

88. Nakanishi, K., Howard, M., Muraguchi, A. et al., Soluble factors involved in B cell differentiation: identification of two distinct T cell-replacing factors (TRF), *J. Immunol.*, 130, 2219, 1983.

89. Julius, M. H., Von Boehmer, H., and Sidman, C., Dissociation of two signals required for activation of resting B cells, *Proc. Natl. Acad. Sci. U.S.A.*, 79, 1989, 1982.

90. Pettersson, S., Pobor, G., and Coutinho, A., MHC restriction of male antigen-specific T helper cells collaborating in antibody response, *Immunogenetics*, 15, 129, 1982.

91. Andersson, J., Schreier, M. H., and Melchers, F., T-cell-dependent B-cell stimulation is H-2 restricted and antigen dependent only at the resting B-cell level, *Proc. Natl. Acad. Sci. U.S.A.*, 77, 1612, 1980.

92. Andersson, J., Bullock, W. W., and Melchers, F., Inhibition of mitogenic stimulation of mouse lymphocytes by anti-mouse immunoglobulin antibodies. I. Mode of actions, *Eur. J. Immunol.*, 4, 715, 1974.

93. Zitron, I. M., Mosier, D. E., and Paul, W. E., The role of surface IgD in the response to thymic-independent antigens, *J. Exp. Med.*, 146, 1707, 1977.

94. Finkelman, F. D., Mond, J. J., Woods, V. L., Wilburn, S. B., Berning, A., Sehgal, E., and Scher, I., Effects of anti-immunoglobulin antibodies on murine B lymphocytes and humoral immune responses, *Immunol. Rev.*, 52, 38, 1980.

95. Primi, D., Lewis, G. K., and Goodman, J. W., The role of immunoglobulin receptors and T cell mediators in B lymphocyte activation. I. B cell activation by anti-immunoglobulin and anti-idiotype reagents, *J. Immunol.*, 125, 1286, 1980.

96. Finkelman, F. D., Scher, I., Mond, J. J., Kessler, S., Kung, J. T., and Metcalf, E. S., Polyclonal activation of the murine immune system by an antibody to IgD. II. Generation of polyclonal antibody production and cells with surface IgG, *J. Immunol.*, 129, 638, 1982.

97. Finkelman, F. D., Muul, L. M., Yaffe, L., Scher, I., Mond, J. J., Kessler, S. W., Ryan, J., Kung, J. T., and Metcalf, E. S., Stimulation of the murine immune system by anti-IgD antibodies: a polyclonal model of B lymphocyte activation by a thymus dependent antigen, *Ann. N.Y. Acad. Sci.*, 399, 316, 1982.

98. Chestnut, R. W. and Grey, H. M., Studies on the capacity of B cells to act as antigen-presenting cells, *J. Immunol.*, 126, 1075, 1981.

99. Chestnut, R. W., Colon, S. M., and Grey, H. W., Antigen presentation by normal B cells, B cell tumors, and macrophages: functional and biochemical comparison, *J. Immunol.*, 128, 1764, 1982.

100. Ryan, J. J., Mond, J. J., Finkelman, F. D., and Scher, I., Enhancement of the mixed lymphocyte reaction by *in vivo* treatment of stimulator spleen cells with anti-IgD antibody, *J. Immunol.*, 130, 2534, 1983.

101. Honjo, T., Immunoglobulin genes, in *Annual Review of Immunology*, Vol. 1, Paul, W. E., Fathman, C. G., and Metzger, H., Eds., Annual Reviews, Palo Alto, Calif., 1983, 499.

102. Tucker, P. W., Chang, H.-L., Richard, J. E., Fitzmaurice, L., Mushinski, J. F., and Blattner, F. R., Genetic aspects of IgD expression. II. Functional implications of the sequence and organization of the Cδ gene, *Ann. N.Y. Acad. Sci. U.S.A.*, 399, 26, 1982.

103. Rowe, D. S. and Fahey, J. L., A new class of human immunoglobulins. II. Normal serum IgD, *J. Exp. Med.*, 121, 185, 1965.

104. Finkelman, F. D., Woods, V., Berning, A., and Scher, I., Demonstration of mouse serum IgD, *J. Immunol.*, 123, 1253, 1979.

105. Maki, R., Roeder, W., Traunecka, A., Sidman, C., Wabl, M. et al., The role of DNA rearrangement and alternative RNA processing in the expression of immunoglobulin delta genes, *Cell*, 24, 353, 1981.

106. Vitetta, E. S., Melcher, U., McWilliams, M., Phillips-Quagliata, J., Lamm, M., and Uhr, J. W., Cell surface Ig. XI. The appearance of an IgD-like molecule on murine lymphoid cells during ontogeny, *J. Exp. Med.*, 141, 206, 1975.

107. Havran, W. L., DiGiusto, D. L., and Cambier, J. C., mIgM:mIgD ratios on B cells: mean mIgD expression exceeds mIgM by 10-fold on most splenic B cells, *J. Immunol.*, 132, 1712, 1984.

108. Bourgois, A., Kitajima, K., Hunter, I. R., and Askonas, B. A., Surface immunoglobulins of lipopolysaccharide-stimulated spleen cells. The behavior of IgM, IgD and IgG, *Eur. J. Immunol.*, 7, 151, 1977.

109. Kanowith-Klein, S., Vitetta, E. S., and Ashman, R. F., The isotype cycle: successive changes in surface immunoglobulin classes expressed by the antigen-binding B cell population during the primary *in vivo* immune response, *Cell. Immunol.*, 62, 377, 1981.

110. Monroe, J. G. and Cambier, J. C., B lymphocyte activation: entry into cell cycle is accompanied by decreased expression of IgD but not IgM, *Eur. J. Immunol.*, 13, 208, 1983.

111. Yuan, D., Regulation of IgM and IgD synthesis in B lymphocytes. II. Translational and post-translational events, *J. Immunol.*, 132, 1566, 1984.

112. Tucker, P. W., Liu, C.-P., Mushinski, J. F., and Blattner, F. R., Mouse immunoglobulin D: messenger RNA and genomic DNA sequences, *Science*, 209, 1353, 1980.

113. Lin, L. and Putnam, F. W., Structural studies of human IgD: isolation by a two-step purification procedure and characterization by chemical and enzymatic fragmentation, *Proc. Natl. Acad. Sci. U.S.A.*, 76, 6572, 1979.

114. Pure, E. and Vitetta, E. S., The murine B cell response to TNP-polyacrylamide beads: the relationship between the epitope density of the antigen and the requirements for T cell help and surface IgD, *J. Immunol.*, 125, 420, 1980.

115. Vitetta, E. S. and Uhr, J. W., Cell surface immunoglobulin. XIX. Susceptibility of IgD and IgM on murine splenocytes to cleavage by papain, *J. Immunol.*, 117, 1579, 1976.

116. Kehry, M., Sibley, C., Fuhrman, J., Schilling, J., and Hood, L. E., Amino acid sequence of a mouse immunoglobulin μ chain, *Proc. Natl. Acad. Sci. U.S.A.*, 76, 2932, 1979.

117. Gray, D., MacLennan, I. C. M., Bazin, H., and Khan, M., Migrant, μ⁺δ⁺ and static μ⁺δ⁻ B lymphocyte subsets, *Eur. J. Immunol.*, 12, 564, 1982.

118. Metcalf, E. S., Mond, J. J., and Finkelman, F. D., Effects of neonatal anti-δ antibody treatment on the murine immune system. II. Functional capacity of a stable sIgM⁺IgD⁻ B cell population, *J. Immunol.*, 131, 601, 1983.

119. Muul, L. M., Mond, J. J., and Finkelman, F. D., Polyclonal activation of the murine immune system by an antibody to IgD. III. Ontogeny, *Eur. J. Immunol.*, 13, 900, 1983.

120. Finkelman, F. D., Mond, J. J., and Metcalf, E. S., Effects of neonatal anti-δ treatment on the murine immune system. I. Suppression of development of surface IgD⁺ B cells and expansion of a surface IgM⁺IgD⁻ B lymphocyte population, *J. Immunol.*, 131, 593, 1983.

121. Skelly, R. B., Baine, Y., Ahmed, A., Xue, B., and Thorbecke, G. J., Cell surface phenotype of lymphoid cells from normal mice and mice treated with monoclonal anti-IgD from birth, *J. Immunol.*, 130, 15, 1983.

122. Jacobson, E. B., Baine, Y., Chen, Y. W., Flotte, T., O'Neil, M. J., Pernis, B., Siskind, G. W., Thorbecke, G. J., and Tonda, P., Physiology of IgD. I. Compensatory phenomena in B lymphocyte activation in mice treated with anti-IgD antibodies, *J. Exp. Med.*, 154, 318, 1981.

123. Hunter, K. W., Finkelman, F. D., and Smith, L. P., Murine malaria: modulation of the immune response by antibodies to IgD, *Fed. Proc. Fed. Am. Soc. Exp. Biol.*, 40, 1071, 1981.

124. Metcalf, E. S., Mond, J. J., and Finkelman, F. D., Abnormal B cell function in mice treated with anti-IgD from birth, *Fed. Proc. Fed. Am. Soc. Exp. Biol.*, 41, 424, 1982.

125. Buck, L. B., Yuan, D., and Vitetta, E. S., A dichotomy between the expression of IgD on B cells and its requirement for triggering such cells with two T-independent antigens, *J. Exp. Med.*, 149, 987, 1979.

126. McFadden, S. F. and Vitetta, E. S., sIgD-negative B cells from neonatal mice do not respond to the thymus-independent antigen TNP-BA in limiting dilution cultures, *J. Immunol.*, 132, 1717, 1984.

127. Cambier, J. C., Vitetta, E. S., Kettman, J. R., Wetzel, G., and Uhr, J. W., B cell tolerance. III. Effects of papain-mediated cleavage of cell surface IgD on tolerance susceptibility of murine B cells, *J. Exp. Med.*, 146, 107, 1977.

128. Vitetta, E. S., Cambier, J. C., Ligler, F. S., Kettman, J. R., and Uhr, J. W., B-cell tolerance. IV. Differential role of surface IgM and IgD in determining tolerance susceptibility of murine B cells, *J. Exp. Med.*, 146, 1977, 1977.

129. Scott, D. W., Layton, J. E., and Nossal, G. J. V., Role of IgD in the immune response and tolerance. 1. Anti-δ pretreatment facilitates tolerance induction in adult B cells *in vitro, J. Exp. Med.*, 146, 1977, 1977.

130. Cambier, J. C., Ligler, F. S., Uhr, J. W., Kettman, J. R., and Vitetta, E. S., Blocking of primary *in vitro* antibody responses to thymus-independent antigens with antiserum specific for IgM or IgD, *Proc. Natl. Acad. Sci. U.S.A.*, 75, 432, 1978.

131. Parkhouse, R. M. E. and Cooper, M. D., A model for the differentiation of B lymphocytes with implications for the biological role of IgD, *Immunol. Rev.*, 37, 105, 1977.

132. Finkelman, F. D., Smith, J. M., and Villacreses, N., The role of antigen specific T help in the *in vivo* generation of a specific antibody response, submitted for publication.

133. Sidman, C. L., Paige, C. J., and Schreier, M. H., B cell maturation factor (BMF): a lymphokine or family of lymphokines promoting the maturation of B lymphocytes, *J. Immunol.*, 132, 209, 1984.

134. Takatsu, K., Tanaka, K., Tominaga, A., Kumahara, Y., and Hamaoka, T., Antigen-induced T cell replacing factor (TRF). III. Establishment of T cell hybrid clone continuously producing TRF and functional analysis of released TRF, *J. Immunol.*, 125, 2646, 1980.

135. Swain, S. L., Dennert, G., Warner, J. F., and Dutton, R. W., Culture supernatants of a stimulated T cell line have helper activity that synergizes with IL2 in the response of B cells to antigen, *Proc. Natl. Acad. Sci. U.S.A.*, 78, 2517, 1981.

136. Isakson, P. C., Pure, E., Vitetta, E. S., and Krammer, P. H., T cell-derived B cell differentiation factor(s). Effect on the isotype switch of murine B cells, *J. Exp. Med.*, 155, 734, 1982.

137. Clayberger, C., Dekruyff, R. H., Fay, R., Pavlaky, M., and Cantor, H., Immunoregulation of T-dependent responses by a cloned dendritic cell, *J. Immunol.*, 133, 1174, 1984.

138. Goodman, M. G. and Weigle, W. O., Activation of lymphocytes by a thiol-derivatized nucleoside: characterization of cellular parameters and responsive subpopulation, *J. Immunol.*, 130, 551, 1983.

139. Finkelman, F., Malek, T., Shevach, E., and Mond, J., B cell IL2 receptor (IL2R) expression *in vivo* and *in vitro, Fed. Proc. Fed. Am. Soc. Exp. Biol.*, 44, 1533, 1985.

140. Finkelman, F., Unpublished data.

141. Mond, J. J., Unpublished observation.

142. DeFranco, A., Personal communication.

143. Mond, J. J., Unpublished data.

144. Vitetta, E., Personal communication.

145. Goroff, D., Unpublished data.

146. Sieckmann, D., Personal communication.

147. Finkelman, F. and Mond, J., Unpublished data.

148. Weber, R., Manuscript in preparation.

149. Finkelman, F., Mond, J., Malek, R., and Shevach, E., Manuscript in preparation.

150. Nakanishi, K., Personal communication.

151. Zubler, R., Personal communication.

Chapter 4

HELPER SIGNALS IN B LYMPHOCYTE DIFFERENTIATION

Neal W. Roehm, H. James Leibson, Albert Zlotnik, Phillipa Marrack,
and John W. Kappler

TABLE OF CONTENTS

I. INTRODUCTION

For many years our laboratory has been interested in characterizing the nature of the helper signals required by B cells in the generation of specific antibody responses. Based on a variety of experimental approaches we have proposed that antigen-primed T cells mediate two types of helper activities in the generation of B cell antibody responses.[1-6] One of these activities was first detected in our hands by the ability of antigen-primed T cells, upon specific restimulation, to nonspecifically drive a bystander antibody response to erythrocyte-associated antigens.[1] Our experimental system for studying the nature of this "antigen-nonspecific" helper activity has been the generation of primary, in vitro plaque-forming cell (PFC) responses to sheep erythrocytes (SRBC).

Using this system it was observed that cell-free culture supernatants of normal spleen cells stimulated with either allogeneic leukocytes or the mitogen Concanavalin A (spleen Con A SN), as well as supernatants of antigen-primed T cells following specific restimulation, contained nonspecific factors which were able to drive the generation of erythrocyte-specific PFC responses by T cell-depleted splenic B cells stimulated with SRBC.[6] The generation of such nonspecific factor preparations was shown to be dependent on the presence of activated T cells and adherent cells (presumably, macrophages [MØ]). In addition, extensive absorption of these preparations with SRBC failed to remove the activity, confirming their antigen-nonspecific nature.

While nonspecific factor preparations, such as spleen Con A SN, were sufficient to drive the antihapten PFC response to SRBC conjugated with trinitrophenol (TNP), they were found to be insufficient in the PFC response stimulated by TNP conjugated to soluble protein antigens such as human gamma globulin (HGG) or keyhole limpet hemocyanin (KLH).[1-5] In addition to nonspecific factors, the PFC response to protein-bound haptens required a second type of helper T cell activity, referred to as "antigen-specific" help, which somehow involved the action of the antigen-specific/major histocompatibility complex (MHC) restricted T cell receptor.

Based on a series of experiments utilizing limiting dilution, differences in the antigen dose-response profile for priming and challenge, as well as differences in antigen fine specificity, we proposed that the two types of helper activities might be mediated by different subsets of T cells.[2,3,7-10] This conclusion received further support from the observation that an I region-specific alloantisera, raised by immunization with mitogen-induced T cell blasts, eliminated a population of helper T cells whose function could be replaced by spleen Con A SN.[4,11] Since the spleen Con A SN was insufficient alone, this result again suggested that at least two types of helper T cell activities were involved at the effector phase in the generation of PFC responses to protein-bound haptens.

In this chapter we will review our studies characterizing the nature of these two types of helper activities and the roles they may play in B cell activation. The focus of our approach in these studies has been the development of an extensive panel of T cell hybridomas. These hybridomas offer a number of advantages for such studies, including rapid growth in the absence of filler cells and exogenous factors. This feature of the hybridomas has served two purposes. First, it has allowed us to test the synergistic effects of nonspecific factors and antigen-specific help, provided by the hybridomas, under conditions where we are confident that the effects of the factors were on the B cells themselves. Second, it has readily allowed the clonal production of lymphokines on a large scale. While to a certain extent somatic cell hybrids suffer from the disadvantage of functional instability due to chromosome loss, we have not found this to be a problem, and in certain cases discussed below, this fact has actually proven advantageous.

II. NONSPECIFIC FACTORS AND THE GENERATION OF B CELL ANTIBODY RESPONSES

A. Interleukin-2 and T Cell Contamination

As part of our effort to characterize the nature of the nonspecific factors present in spleen Con A SN, we began the production of a panel of T cell hybridomas. This approach was based on the idea that as a clonal source, T cell hybridomas might produce a more restricted set of factors than would be expected following polyclonal stimulation of unfractionated leukocytes.

The first of these hybridomas, FS6-14.13, was produced by the polyethylene glycol-induced fusion of normal splenic T cells with the azaguanine-resistant T cell tumor BW5147 and selected by the ability of a Con A-induced culture supernatant from the hybridoma (FS6 Con A SN) to drive the generation of a primary antierythrocyte PFC response by T cell-depleted splenic B cells stimulated with SRBC.[12] Subsequent analysis of FS6 Con A SN revealed the presence of interleukin-2 (IL-2) as defined by the supernatant's ability to maintain the continued growth and viability of the IL-2-dependent T cell line HT-2 in a T cell growth factor assay. This fact raised the possibility that at least one component of the B cell helper activity of FS6 Con A SN was due to the indirect effects of IL-2, in combination with the Con A used to induce its production, on contaminating T cells in the B cell preparation.

In an attempt to test this possibility, more rigorous methods were introduced for the depletion of T cells. The protocol that was developed involved the treatment of donor mice with rabbit antimouse thymocyte serum in vivo to deplete recirculating T cells, followed by negative selection in vitro with a cocktail of antibodies and complement.[12-14] In its current form this cocktail includes: B cell-absorbed rabbit antimouse thymocyte serum, T24/40.7 anti-Thy-1 monoclonal antibody (Dr. Ian Trowbridge, Salk Institute, La Jolla, Calif.), MK 2.2 anti-"Qa-like" monoclonal antibody, B16/146 anti-Qat-4 monoclonal antibody (Dr. Ulrich Hammerling, Sloan-Kettering Memorial Cancer Institute, New York), ADH4 (15) anti-Lyt-2.2 monoclonal antibody (Dr. Paul Gottlieb, University of Texas, Austin), and GK 1.5 anti-L3T4 monoclonal antibody (Drs. Deno Dialynas and Frank Fitch, The University of Chicago). This method of B cell preparation was further supplemented by the addition of α-methyl-D-mannoside to the cultures to prevent effects due to residual Con A in the nonspecific factor preparations.

Use of this rigorous method of T cell depletion abrogated the ability of FS6 Con A SN to directly drive the generation of in vitro anti-SRBC PFC responses, while leaving intact the response driven by spleen Con A SN.[13] As an indication of the sensitivity of the B cell response to T cell contamination in the presence of IL-2, the addition of less than 1% normal spleen cells to the T cell-depleted cultures allowed FS6 Con A SN to elicit significant responses. These results indicated that the ability of IL-2 to be a sufficient helper factor alone was dependent on the presence of contaminating T cells and that spleen Con A SN contained other B cell helper factors, in addition to IL-2. While these results indicated that IL-2 alone was not sufficient for a B cell response, they did not address the question of whether IL-2 could act directly as a B cell helper factor in concert with other interleukins. The fact that spleen Con A SN stimulated PFC responses even in thoroughly T cell-depleted B cells suggested that factors other than IL-2 might be involved. If this hypothesis were correct, then removal of IL-2 from spleen Con A SN should not affect its helper activity.

B. At Least Two Factors Are Required in the PFC Response to Sheep Erythrocytes

To test this prediction, we removed IL-2 from spleen Con A SN by incubation with activated T cells.[13] This method is based on the fact that following mitogen or antigen-specific/MHC-restricted activation, T cells are induced to express receptors for IL-2.[15] Subsequent binding of IL-2 to this specific receptor leads to internalization and lysosomal degradation of the IL-2 by the T cell. Thus, incubation of spleen Con A SN with mitogen-activated T cell blasts or the IL-2-dependent T cell line HT-2 abrogated virtually all helper activity in the generation of anti-SRBC PFC responses, in parallel with the loss of T cell growth factor activity. In both these cases the addition of FS6 Con A SN, which was unable to support the anti-SRBC PFC response alone, restored the helper activity of the IL-2-depleted Con A SN. The specificity of the IL-2 depletion was confirmed by the failure of absorptions with normal spleen cells, lipopolysaccharide (LPS)-activated B cells, MOPC-21, or BW5147 to affect the B cell helper activity of spleen Con A SN.

These results indicated that at least two helper factors were required in the B cell PFC response to SRBC. One of these factors was removed by incubation with IL-2-dependent T cells, while the other(s) was not. The helper activity which was removed by activated T cells was present in FS6 Con A SN, while the nonabsorbable activity was not. For lack of a better term we have referred to this latter activity as IL-X. We found that the most reliable source of this activity was spleen Con A SN harvested at day 4 of culture.[13,14] While IL-2 is readily detectable at 24 hr of culture, by day 4 of culture utilization of IL-2 by the mitogen-activated T cells exceeds production and the levels of IL-2 fall below the level of detectability in the HT-2 T cell growth factor assay. The active component(s) of the 4-day spleen Con A SN were found to fall within the 30,000- to 50,000-dalton range upon Sephadex® G75 chromatography, and we will refer to nonspecific factors prepared in this manner as IL-X.

In an attempt to strengthen the conclusion that it was IL-2 which was acting as a B cell helper factor, two additional experiments were performed.[13] In the first we compared the Con A SN from a panel of independent T cell hybridomas for both IL-2, using the T cell growth factor assay, and for their ability to restore the PFC response of T cell-depleted splenic B cells stimulated with SRBC in the presence of IL-2-depleted spleen Con A SN. Calculation of the units of activity of these Con A SN from titration curves obtained in these two assays established that the levels of B cell helper activity were directly correlated with the levels of IL-2, both qualitatively and quantitatively. In a second experiment, FS6 Con A SN was fractionated by molecular weight using Sephadex® G100 column chromatography. Analysis of the fractions indicated that both IL-2 and the B cell helper activity, which synergized with IL-2-depleted spleen Con A SN in the anti-SRBC PFC response, coeluted with an apparent molecular weight of approximately 40,000.

These results suggested that IL-2 was capable of directly acting as a B cell helper factor in the generation of specific antibody responses. This conclusion has certainly generated considerable controversy for several reasons. First, although we have taken elaborate measures to deplete T cells from our cultures, we cannot eliminate the possibility that IL-2 functions by indirectly stimulating residual contaminating T cells. Second, while IL-2 is readily depleted by incubation with activated T cells, attempts to deplete IL-2 by incubation with activated B cells have been unsuccessful. Third, studies using internally radiolabeled, purified IL-2 failed to detect significant binding to LPS-activated B cells, while such binding was readily detectable with activated T cells.[15] However, in more recent studies, monoclonal antibodies have been developed which appear to be directed against the IL-2 receptor complex.[16] Using these antibodies it has been possible to detect the presence of the IL-2 receptor on LPS-activated B cells, albeit at levels tenfold lower than on activated T cells. There is, at this time, no irrefutable evidence that IL-2 can directly act as a B cell helper factor, however, in our hands, no combination of helper factors has replaced the requirement for IL-2 in the generation of B cell antibody responses.

C. Nonspecific Factors and PFC Response to "Thymus-Independent" Antigens

A number of studies published during this period indicated that rigorous T cell depletion of murine spleen cells could reveal previously unrecognized effects of T cells on B cell responses to putative thymus-independent (TI) antigens. In preliminary studies we also observed that while significant PFC responses could be elicited with TI antigens using spleen cells depleted of T cells by treatment with anti-Thy-1 and complement, such responses were virtually eliminated by rigorous T cell depletion as described above. The resulting implication that T cells were in fact playing a role in antibody responses to TI antigens was further supported by the observation that the addition of spleen Con A SN significantly enhanced the antihapten PFC response of MØ and T cell-depleted spleen cells stimulated with a variety of TI antigens including: DNP-ficoll, TNP-*Brucella abortus,* and TNP-LPS. The enhancement of these PFC responses by spleen Con A SN was antigen dependent.[17]

The effects of FS6 Con A SN and IL-X preparations on the antibody response to TI antigens were also examined.[17] The addition of FS6 Con A SN alone significantly enhanced the PFC response elicited with a variety of TI antigens. This effect of FS6 Con A SN was completely abolished by incubation of the supernatant with IL-2-dependent HT-2 cells, in parallel with the loss of T cell growth factor activity. In contrast, the IL-X preparation was without effect when used alone. However, when IL-X and FS6 Con A SN were used in combination, there was a synergistic effect on the PFC response elicited with the TI antigens, which was equal to or greater than that observed with spleen Con A SN. These results thus established a role for nonspecific factors, including T cell lymphokines, in the antibody response to a variety of putative TI antigens.

D. At Least Three Nonspecific Factors Are Required for Optimal PFC Responses to Sheep Erythrocytes

In the above studies we established a role for FS6 Con A SN and a 30,000- to 50,000-molecular weight factor(s) present in 4-day spleen Con A SN (IL-X), in the PFC response to SRBC by splenic B cells rigorously depleted of T cells. An issue that was not resolved in these studies was the possible role of MØ in this response. Preliminary studies demonstrated that while MØ depletion had little or no effect on the PFC responses elicited with normal spleen Con A SN, MØ depletion by Sephadex® G10 passage significantly reduced the synergistic response elicited with FS6 Con A SN and IL-X.[14] Two hypotheses could be proposed to explain this result: first, that a MØ-produced factor was required for the response, and second, that the effects of the factors present in FS6 Con A SN and/or IL-X were somehow delivered indirectly to the B cell via MØ.

In an attempt to distinguish between these two possibilities we tested a variety of MØ-derived supernatants for their ability to reconstitute the anti-SRBC PFC response of MØ and T cell-depleted B cells, when used in combination with FS6 Con A SN and IL-X. After testing a variety of MØ-derived supernatants for such activity, we found the constitutive culture supernatant of the MØ tumor line P388D1 (P388 SN) to be the most reliable source, with helper activity indistinguishable from that produced by normal MØ.[14] Thus, using MØ and T cell-depleted splenic B cells, we found that the anti-SRBC PFC response elicited with spleen Con A SN could be completely reconstituted by the addition of three nonspecific factor preparations: P388 SN, FS6 Con A SN, and IL-X (4-day spleen Con A SN from which the 30,000- to 50,000-molecular weight components were isolated by Sephadex® G75 column chromatography). The failure of P388 SN to substitute for either of the other factor preparations implied that none of the factors present in FS6 Con A SN or IL-X functioned by indirectly stimulating MØ factor production, but rather that the relevant factors in each preparation operated via direct effects on the B cells.

At this time the identity of the active component(s) of P388 SN in our B cell assays has not been conclusively determined. Historically, supernatants of P388D1 have been used as a reference source for interleukin-1 (IL-1),[18,19] however, we have not always found a direct correlation between the level of helper activity in our SRBC assays and their lymphocyte activating factor (LAF) activity in thymocyte proliferation assays. Sephadex® G75 column fractionation of P388 SN has, however, revealed an active component in the anti-SRBC PFC response corresponding to a molecular weight of approximately 15,000, which would be consistent with a role for IL-1 as suggested by other investigators.[20,21]

In an attempt to confirm that the requirement for each of the three interleukin preparations (P388 SN, FS6 Con A SN, and IL-X) reflected qualitative differences in the factors they contained, a series of titration experiments were performed. Each nonspecific factor preparation was titrated into B cell cultures in the presence of optimal concentrations of the two other preparations. These studies demonstrated that in the presence of saturating concentrations of two of the nonspecific factor preparations, the magnitude of PFC response was directly related to the amount of the third preparation added.

These results indicated that at least three nonspecific helper factors contribute to the activity of spleen Con A SN.[14] This conclusion was also supported by kinetic studies performed to determine the optimal time of addition of each interleukin preparation during the course of the 4-day culture. The optimal time of addition of both the FS6 Con A Sn and the P388 SN was at the initiation of culture, suggesting that the interleukins they contain act very early in B cell activation. In contrast, the optimal time of addition of the IL-X preparation was 24 hr after the initiation of culture. The "late-acting" nature of the IL-X preparation is reminiscent of the activity associated with T cell replacing factor (TRF), however, unlike TRF, IL-X does not totally replace helper T cell activity in anti-SRBC antibody responses.[22]

E. Clonal Source of Interleukin-X

In continuing our studies we attempted to produce a T cell hybridoma for use as a clonal source of IL-X activity free from IL-2.[23] Screening the Con A-induced culture supernatants of 64 independent hybridomas revealed that about 40% produced IL-2, but less than 10% produced the IL-X activity. Of the six IL-X producers, FS7-20 clearly produced much higher levels of activity than the others, but also produced IL-2.

At this point we attempted to take advantage of the fact that upon extended culture somatic cell hybridomas are often karyotypically unstable. Therefore, after extended culture FS7-20 was cloned at limiting dilution. Each of the 29 clones obtained were induced with Con A and the supernatants analyzed for IL-2, IL-X, interferon-gamma (IFN$_\gamma$, assayed by inhibition of viral replication), MØ Ia inducing factor (IaIF), and MØ activating factor (MAF, assayed by induction of peritoneal MØ cytotoxicity against mouse L929 cells).[23] While some of the 29 clones produced either none or all of the lymphokine activities, in the remaining clones their was a striking lack of correlation between IL-2 and the other activities IL-X, IFN$_\gamma$, MAF, and IaIF, whose production was observed to be correlated.

To further examine the reproducibility of these correlations, two of the clones were again cloned by limiting dilution and examined for production of the five lymphokine activities following Con A stimulation.[23] The Con A SN were each titrated in the assays and units of activity per milliliter determined, defining one unit of activity as the supernatant volume required to give 50% of the maximal response which occurred at saturation. Once again, such analyses established a striking correlation between IL-X, IFN$_\gamma$, IaIF, and MAF, both qualitatively and quantitatively, while IL-2 was frequently observed to segregate independently from these four activities.

These data suggested two possibilities. First, that the IL-X, MAF, and IaIF activities were all manifestations of IFN$_\gamma$, or alternatively, these lymphokine activities were mediated by two or more different molecular species which are encoded by linked genes and/or coordinately expressed following Con A activation. In support of the former interpretation each of the four lymphokine activities — IL-X, MAF, IaIF, and IFN$_\gamma$ — was found to be sensitive to pH 2 incubation, while IL-2 activity was unaffected by such treatments. Sensitivity to pH 2 incubation has been a characteristic feature of IFN$_\gamma$ activity in the inhibition of viral replication.

The Con A-induced culture SN of one of the T hybridomas, FS7-20.6.18 (FS7 Con A SN), proved to be a potent source of IL-X activity with little or no IL-2 (<10 units/mℓ, limit of detectability). The B cell helper activity of FS7 Con A SN was indistinguishable from that of IL-X preparations. In the presence of optimal concentrations of P388 SN and FS6 Con A SN, the anti-SRBC PFC response was directly proportional to the amount of FS7 Con A SN added. Kinetic studies demonstrated that like IL-X, FS7 Con A SN was optimally effective when added 24 hr after the initiation of culture.

Thus, using MØ and T cell-depleted B cells, we have been able to demonstrate the synergy of three nonspecific factor preparations in the PFC response to SRBC, the IL-1 containing constitutive culture SN of the MØ tumor line P388D1, the IL-2 containing Con A SN of the T cell hybridoma FS6-14.13, and the IFN$_\gamma$ containing Con A SN of the T cell hybridoma FS7-20.6.18.[23] When the three nonspecific factor preparations were used at optimal concentrations and times, the anti-SRBC PFC response generated was significantly greater than that observed with spleen Con A SN. These results suggest that at least three nonspecific factors contribute to the helper activity of spleen Con A SN in the generation of anti-SRBC PFC responses, and their effects are probably mediated directly on the B cells themselves.

III. PFC RESPONSES TO SOLUBLE PROTEIN ANTIGENS

A. Requirement for Antigen-Specific T Cell Help

While the nonspecific helper factors present in the Con A-induced culture supernatant of normal spleen cells appear sufficient for the elicitation of PFC responses to erythrocyte-associated antigens, this has not been true for PFC responses to soluble protein antigens. This fact has made it difficult to reconcile data regarding the effects of nonspecific helper factors and the effects of antigen-specific helper T cells. A possible resolution of this dichotomy is that while nonspecific factors play a necessary role in the generation of antibody responses to soluble protein antigens, additional "antigen-specific" helper T cell signals are also required.

Attempts to demonstrate the synergy between antigen-specific helper T cells and nonspecific factors have been extremely difficult, because the factors may have effects on helper T cell function instead of or in addition to their effects on the B cells themselves. In an attempt to circumvent this problem we tested a panel of T cell hybridomas for their ability to deliver an "antigen-specific" helper signal in the anti-TNP PFC response induced by protein-bound haptens.[5] Unlike the T cell hybridomas described above whose antigen/MHC specificity is unknown, the hybridomas used in these studies were specifically selected for their ability to produce IL-2 following antigen-specific/MHC restricted activation.[24] To test the helper activity of these T hybridomas, we used a system previously developed to measure the activity of normal carrier-primed T cells, in which the cells to be assayed are titrated into Mishell-Dutton cultures of T cell-depleted splenic B cell/MØ, stimulated with TNP-coupled protein antigen. Because we were specifically interested in the "antigen-specific" helper activity of the T cell hybridomas under conditions where nonspecific helper factors would not be limiting, the cells were tested for help in the presence and absence of spleen Con A SN.

Because these hybridomas grow constitutively in the absence of filler cells or exogenous growth factors, and because receptor-mediated activation is known to be factor independent,[25] we are confident the effects of the nonspecific helper factors are on the B cells themselves and not on the T hybridomas.

The antigen-specific/Ia restricted T cell hybridomas proved to be potent sources of "antigen-specific" helper activity in the generation of anti-TNP PFC responses to protein-bound haptens.[5] Delivery of the helper signal by the T hybridomas was shown to be specific for the protein antigen used as the hapten carrier. For example, using two hybridomas, AODH 7.1 (HGG/I-E^d) and AODK 10.4 (KLH/I-A^d), as a reciprocal pair, the HGG specific hybridoma AODH 7.1 drove the antihapten PFC response stimulated by TNP-HGG, but not by TNP conjugated to an irrelevant carrier KLH. Conversely, the KLH-specific hybridoma, AODK 10.4, drove the PFC response stimulated by TNP-KLH, but not by TNP-HGG.

Two features of the helper signal delivered by the antigen-specific/Ia restricted T hybridomas clearly distinguished their "antigen-specific" helper activity from the production of nonspecific factors.[5] The first of these features was the requirement for linked recognition of the hapten-carrier. Linked recognition is a characteristic that has been associated with carrier-specific help in the generation of antihapten antibody responses, in which collaboration between hapten-primed B cells and carrier-primed T cells does not commonly occur upon stimulation with free carrier and hapten conjugated to an irrelevant carrier.[26] As an example from our studies, the HGG-specific T hybridoma AODH 7.1 was able to drive the anti-TNP PFC response of B cells stimulated with TNP conjugated to HGG, but not with TNP conjugated to an irrelevant carrier KLH, even in the presence of free HGG at concentrations up to 1 mg/mℓ. Control studies demonstrated that the hybridoma AODH 7.1 was able to respond to the free HGG presented by the B cell/MØ, leading to the production of IL-2.

A second feature of the helper activity of these hybridomas was that delivery of the signal was restricted by H-2 antigens on the responding hapten-specific B cells.[5] This requirement was demonstrated in an experiment using two KLH-specific hybridomas, AODK 10.4, which is I-A^d restricted, and BDK 11.1, which is I-A^b restricted, as a reciprocal pair. The I-A^d restricted hybridoma AODK 10.4 drove the antihapten PFC response of H-2^d B10.D2 B cells, but not the response of H-2^b B10 B cells, even in the presence of functional irradiated H-2^d antigen-presenting cells. Conversely, the H-2^b restricted hybridoma BDK 11.1 drove the PFC response of H-2^b B10 B cells, but not H-2^d B10.D2 B cells, even in the presence of functional, irradiated H-2^b antigen-presenting cells. Control experiments demonstrated that the presence of the irradiated H-2 disparate B cell/MØ used as antigen-presenting cells did not inhibit the PFC response generated by the nonirradiated B cells. Additional controls further demonstrated that the irradiated B cell/MØ were capable of presentation of KLH to the appropriate T cell hybridoma (leading to IL-2 production), but were unable to generate PFC responses. We concluded that delivery of the helper signal by the T cell hybridoma was controlled at least by H-2 antigens on the responding hapten-specific B cells. Subsequent studies using H-2 recombinant strains of mice and Ia-specific monoclonal antibodies have demonstrated that these H-2 encoded restriction elements are the I-A or I-E molecules. The fact that the H-2 compatible antigen-presenting cells were able to activate nonspecific factor production by the T hybridomas, but were unable to circumvent the requirement for an I region-restricted interaction with the responding hapten-specific B cells, clearly distinguished the "antigen-specific" helper activity of the hybridomas from the production of nonspecific factors.

B. Requirement for Nonspecific Factors in the PFC Response to Soluble Protein Antigens

In testing the helper activity of the panel of T hybridomas, we noted that the elicitation of optimal PFC responses to protein-bound hapten was to varying degrees dependent of the addition of spleen Con A SN.[5] Thus, while significant PFC responses were elicited by some hybridomas in the absence of spleen Con A SN, the elicitation of responses with other hybridomas was absolutely dependent on the addition of spleen Con A SN. This result provides further support for the hypothesis that PFC responses to soluble protein antigens requires both antigen-specific help and nonspecific factors. We have interpreted the differential requirements of the PFC response for the addition of spleen Con A SN to reflect different abilities of the T cell hybridomas to secrete the necessary lymphokines. Consistent with this interpretation is the fact that those hybridomas which do not require the addition of spleen Con A SN to drive PFC responses to soluble protein antigens are also able to drive bystander PFC responses to sheep erythrocytes, following antigen-specific/MHC restricted activation.

Using two T cell hybridomas from our panel (AODH 7.1 and AODK 10.4) we have systematically analyzed the nonspecific factor requirements of the anti-TNP response generated by hapten-primed B cells stimulated with protein-bound hapten.[27] While AODH 7.1 was able to directly drive the PFC response of T cell-depleted splenic B cell/M in the absence of spleen Con A SN, these responses were rendered completely dependent on the addition of nonspecific factors when the B cells were depleted of MØ by Sephadex® G10 passage and rigorously T cell depleted by treatment with antithymocyte serum in vivo and with the anti-T cell cocktail and complement in vitro.

Our approach in characterizing the nonspecific factors required in the antihapten PFC response was based on the previous demonstration that a combination of three nonspecific factor preparations could substitute for spleen Con A SN in the primary PFC response to sheep erythrocytes. Figure 1 summarizes the effects of the nonspecific factor preparations on PFC responses to protein-bound hapten using either AODH 7.1 (HGG/I-E^d) or AODK 10.4 (KLH/I-A^d) as sources of antigen-specific help. When each interleukin preparation was added alone, marginal responses were observed. In paired combinations of the nonspecific factor preparations, significant responses were observed, however, optimal responses required the addition of all three preparations: FS6 Con A SN plus FS7 Con A SN plus P388 SN. These results directly parallel the nonspecific factor requirements of the PFC response to sheep erythrocytes.

It was particularly noteworthy that no combination of nonspecific factors replaced the requirement for an antigen-specific helper signal provided by the T hybridomas. As shown in Figure 2, in the presence of saturating concentrations of each of the three nonspecific factor preparations, the antihapten PFC response was linearly related to the number of T hybridoma cells added, and no responses were observed in their absence. Subsequent studies demonstrated that in the presence of the three interleukin preparations, delivery of the helper signal by the hybridomas demonstrated the two classic features of antigen-specific help, namely, the requirement for linked recognition of hapten-carrier and H-2 (I-A or I-E) restricted interaction with the responding hapten-specific B cells.[27]

The generation of optimal antihapten PFC responses required the addition of all three nonspecific factor preparations (P388 SN, FS6 Con A SN, and FS7 Con A SN), and in combination their activity was substantially greater than that of spleen Con A SN.[27] As in the SRBC system the requirement for each of the three interleukin preparations for optimal antihapten PFC responses appeared to reflect qualitative differences in the factors they contain. Thus, in the presence of optimal concentrations of the antigen-specific/Ia restricted helper T hybridoma and two of the nonspecific factor preparations, the PFC response was directly related to the amount of the third nonspecific factor preparation added.

With regard to the kinetics of their effects, the elicitation of optimal antihapten PFC responses required that the antigen-specific/Ia restricted T hybridoma, P388 SN, and

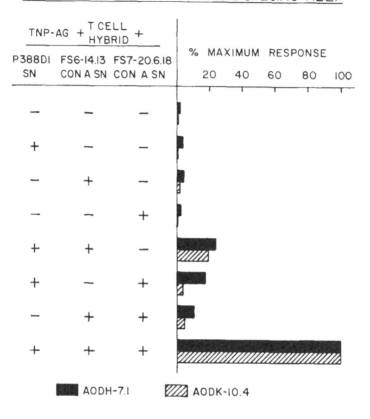

FIGURE 1.   Ability of various combinations of the nonspecific factor preparations to drive an antigen-specific helper T cell-dependent PFC response to protein-bound hapten. Hapten-primed, T cell and MØ-depleted B10.D2 B cells were
stimulated with TNP-HGG in presence of optimal concentrations of AODH 7.1
(HGG/I-E^d) hybridoma cells or TNP-KLH in the presence of AODK 10.4
(KLH/I-A^d) hybridoma cells. The indicated helper factor preparations were
added at optimal concentrations and times. On day 4 of incubation the anti-
TNP PFC/culture was determined and expressed as a percentage of the maximal
response.

FS6 Con A SN all be added at the initiation of culture; the longer the delay in their
addition the lower the response observed on day 4. In contrast, FS7 Con S SN was
optimally effective when added 24 hr after the initiation of culture, implying the relevant factors it contained were "late acting". The kinetics of these effects of the nonspecific factor preparations directly parallel those observed in the anti-SRBC PFC responses.

While the nature of the factors present in each preparation which are active in the
PFC response to soluble protein antigens remains to be conclusively determined, these
data do demonstrate that at least three interleukins, in addition to an antigen-specific/
Ia restricted helper signal, are required for an optimal response. By analogy to the
generation of anti-SRBC PFC responses by MØ and T cell-depleted B cells, we are
confident that the effects of the nonspecific factors on the antibody response reflect
their direct effects on the B cells themselves.

Two types of experiments have been performed to determine the identity of the T
cell lymphokines which are active in the antihapten PFC response.[27] In the first, FS7

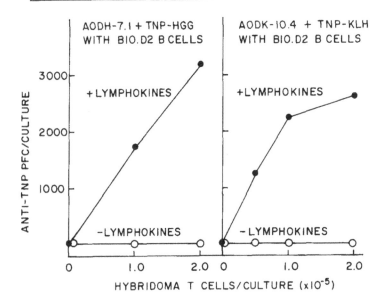

FIGURE 2.   Titration of antigen-specific/MHC-restricted helper T cell hybridomas AODH 7.1 (HGG/I-E^d) or AODK 10.4 (KLH/I-A^d) in the absence or presence of FS6 Con A SN plus FS7 Con A SN plus P388 SN, all added at optimal concentrations and times. The anti-TNP PFC/culture was determined on day 4 of incubation.

Con A SN was subject to pH 2 incubation and its B cell helper activity compared with that of an appropriate control. pH 2 incubation of FS7 Con A SN abrogated its helper activity, consistent with the conclusion that IFN-γ was an active component of the preparation in this assay.

In a second experiment, we attempted to deplete the helper activity of FS6 Con A SN by incubation with the IL-2-dependent T cell line HT-2. IL-2 depletion of FS6 Con A SN significantly reduced, but did not abrogate, its helper activity. The fact that the B cell helper activity of the HT-2-absorbed FS6 Con A SN was significantly greater than would be expected, based on the level of residual T cell growth factor activity, suggests that FS6 Con A SN may contain additional factors which are active in the PFC response to protein-bound hapten. Consistent with this possibility is the fact that FS6 Con A SN has been demonstrated to contain a second lymphokine activity, B cell growth factor (BCGF).[28,29] In our hands, this lymphokine activity has been associated with a 15- to 20-kdalton molecule which synergizes with rabbit F(ab')$_2$ antimouse immunoglobulin in the induction of proliferation by purified B cells cultured at low cell density.[30]

The HT-2-absorbed FS6 Con A SN used in the above studies was virtually devoid of IL-2 activity, but retained approximately 50% of its BCGF activity.[27] Nevertheless, it was unable to replace the activity of FS6 Con A SN at any concentration, again implying a role for IL-2. In addition, when tested in the anti-SRBC PFC response, this IL-2-depleted preparation of FS6 Con A SN was without activity, consistent with our inability to demonstrate any role for BCGF in this response. However, when tested in the PFC response to protein-bound hapten, this preparation did have some residual activity. Whether this reflects an effect of BCGF remains to be determined, but such a possibility would be consistent with data discussed below.

IV. CLONED LYMPHOKINES AND THE PFC RESPONSE TO SHEEP ERYTHROCYTES

With the recent advances in recombinant DNA technology, it has become possible to isolate cDNA clones encoding lymphokine mRNA and obtain their functional expression in *Escherichia coli.* These techniques have led to the availability of several T cell lymphokines with activities thousands of times greater than conventional sources and free from other interleukins. We have been able to obtain cloned murine IFN$_\gamma$ and cloned human IL-2 and have examined their effects on the generation of primary anti-SRBC PFC responses.

Purified, LPS-free, cloned murine IFN$_\gamma$ prepared in *E. coli* was kindly provided to us by Genentech, Inc. (San Francisco, Calif.). This preparation of IFN$_\gamma$ was compared with FS7 Con A SN for its ability to synergize with FS6 Con A SN and P388 SN in the generation of anti-SRBC PFC by MØ and T cell-depleted splenic B cells.[30,31] Titration of these two factor preparations demonstrated that supernatants having equal numbers of units of IFN$_\gamma$ (based on inhibition of viral replication) had indistinguishable helper activity in the generation of anti-SRBC PFC. In addition, kinetic studies demonstrated that both preparations were optimally effective when added on day 1 of the 4-day response. Based on these results we would conclude that IFN$_\gamma$ is a late-acting B cell helper factor in the generation of specific antibody responses. Preliminary studies have indicated that cloned IFN$_\gamma$ will also synergize with FS6 Con A SN and P388 SN in the antigen-specific helper T cell-dependent PFC response to protein-bound haptens. In our hands IFN$_\gamma$ is the only lymphokine we have observed which shares with T cell-replacing factor the property of being a late-acting factor in B cell responses. Whether, in fact, other interleukins have this property remains to be demonstrated.

We have also examined the B cell helper activity of purified, cloned human IL-2 prepared in *E. coli* and kindly provided to us by David Mark of Cetus (Emeryville, Calif.). Preparations of cloned IL-2 and FS6 Con A SN were titrated into cultures of T cell and MØ-depleted splenic B cells, stimulated with SRBC in the presence of P388 SN and cloned IFN$_\gamma$. Both preparations had equivalent activity in the generation of anti-SRBC PFC when compared on the basis of the units of T cell growth factor activity they contained, determined using the IL-2-dependent T cell line HT-2. This result demonstrates that the B cell helper activity of FS6 Con A SN in the anti-SRBC PFC response can be completely attributed to IL-2. Further studies will be necessary to discern the mechanism of the effect of IL-2 on the B cell response.

V. INTERLEUKIN-INDUCED INCREASE IN B CELL Ia EXPRESSION

Many studies, including our own, have investigated the role of interleukins in B cell activation. In virtually every system studied to date, nonspecific factors have been without apparent effects in the absence of additional activating signals provided by mitogens, antiimmunoglobulin antibodies, or antigen plus Ia restricted helper T cell signals. These observations have suggested models in which these activating signals induce or activate receptors for interleukins, rendering the cell sensitive to their effects. Such a model would closely parallel studies demonstrating the ability of mitogens or antigen/MHC activation of T cells to induce receptors for IL-2.[15] In fact, this system might be directly applicable to B cells, based on evidence demonstrating that a monoclonal antibody directed against the putative IL-2 receptor complex can be used to detect its presence on the surface of LPS-activated B cells, but not on resting cells.[16]

Until recently, these models of B cell activation would have been completely consistent with our own observations. However, we now have reason to speculate that normal resting B cells express receptors for at least some nonspecific factors, which can have direct effects on B cells in the absence of other ancillary signals.

The basis for this conclusion was derived from the observation that nonspecific factors present in P388 SN and FS6 Con A SN were able to directly induce a dose-dependent increase in Ia expression by populations of normal splenic B cells, depleted of MØ and T cells as previously described.[32] Relative Ia expression was determined by immunofluorescent staining with monoclonal antibodies and flow cytometric analysis. A representative histogram depicting the changes in Ia expression induced by the nonspecific factor preparations is shown in Figure 3. In a series of experiments comparing relative mean fluorescence to that of the control population, P388 SN induced a 4.9 ± 0.9-fold increase in mean Ia expression and FS6 Con A SN 10.7 ± 1.5-fold increase. When used in combination at subsaturating concentrations, the effects of P388 SN and FS6 Con A SN were additive at best and never synergistic.

Several points can be made with respect to these observations. First, induction of increased B cell Ia expression by the nonspecific factor preparations was extremely rapid, reaching maximum levels within 24 hr of culture.[32] Second, comparison of the curves in Figure 3 reveals that there was only a small overlap in the profiles of the control population and cells incubated with FS6 Con A SN. This result suggests that virtually all the cells were stimulated to increased Ia expression and not just a subpopulation. This conclusion is also supported by the fact that greater than 90% of the cells recovered after culture were viable and thus the increase in Ia expression does not reflect the selective survival of a subset of cells with high Ia expression. Third, analysis of cell cycle state by acridine orange staining demonstrated that greater than 95% of the control B cells cultured in the absence of factors were in G_0, suggesting that the interleukins were able to directly induce increased Ia expression by resting B cells in the apparent absence of other activating signals.[32] These effects of the interleukins on B cell Ia expression are similar in kinetics and magnitude to those previously observed with antiimmunoglobulin antibodies.

In an attempt to determine whether the induction of increased Ia expression, as determined by immunofluorescence, might influence T-B interactions, we tested the ability of the B cells to present antigen to a panel of antigen-specific/MHC restricted T cell hybridomas.[32] Successful presentation of antigen was monitored by the production of IL-2 by the T hybridomas. A representative experiment is shown in Figure 4. Purified B cells were cultured either alone or in the presence of FS6 Con A SN plus P388 SN. After 24 hr of incubation the B cells were titrated into cultures containing ovalbumin (OVA) as antigen and the T hybridoma 3DO-54.8 (OVA/I-A^d). After an additional 24 hr of culture the supernatants were tested for the presence of IL-2. As shown in Figure 4, the induction of increased Ia expression was correlated with an increased ability of the B cells to present antigen to the T hybridomas in an I-A restricted manner.

Clearly, the interleukins could have induced a number of changes which might have influenced the ability of the B cells to present antigen. In an attempt to more closely focus on interleukin-induced changes occurring at the cell surface, we took advantage of the recent demonstration that glutaraldehyde-fixed accessory cells are capable of MHC-restricted presentation of antigens which have been processed in vitro by chemical or enzymatic degradation.[33] Thus, B cells which had been induced to increased Ia expression by incubation with FS6 Con A SN plus P388 SN, and then fixed with glutaraldehyde, demonstrated an enhanced ability to present trypsin-digested ovalbumin to T hybridomas of appropriate specificity, as compared to glutaraldehyde-fixed control cells which had been cultured in the absence of the interleukin preparations.[32] While this experiment does not conclusively demonstrate that the enhanced antigen-presenting function was due to increased Ia expression, these results were consistent with this possibility.

While we have not been able to demonstrate a role for BCGF in the PFC response

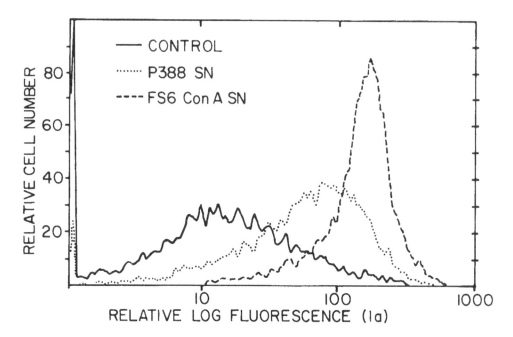

FIGURE 3.   Interleukin-induced increase in B cell Ia expression. T cell and MØ-depleted splenic B cells were cultured for 24 hr either alone or in the presence of P388 SN or FS6 Con A SN. Cells were stained with affinity purified biotin-conjugated anti-I-A^{b,d} monoclonal antibody followed by fluorescein-conjugated avidin and analyzed by flow cytometry. Five thousand cells were analyzed for each histogram. Relative fluorescence (mean channel number) of control cells was 12.0, cells incubated with P388 SN 50.1, and with FS6 Con A SN 129.1.

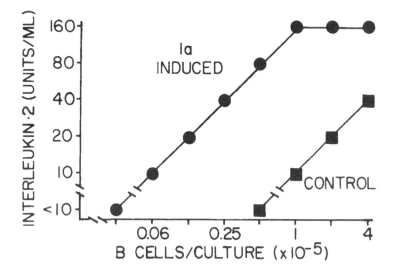

FIGURE 4.   Enhanced antigen presenting ability of normal B cells incubated with interleukin preparations. T cell and MØ-depleted splenic B cells were cultured for 24 hr either alone (control) or with P388 SN plus FS6 Con A Sn (Ia induced). When compared to the control the interleukin preparations induced a 6.3× increase in relative fluorescence (Ia) of the B cells. The two populations were then titrated for their ability to present OVA to the T hybridoma 3DO-54.8 (OVA/I-A^d). Presentation of antigen was determined by the production of IL-2 by the T hybridoma.

to sheep erythrocytes,[30] data previously discussed suggested that a factor present in FS6 Con A SN, which was not depleted by incubation with IL-2-dependent T cells, might be involved in the PFC response to protein-bound hapten. One might speculate that since these responses require a B cell Ia-restricted helper T cell signal, the lymphokine present in FS6 Con A SN which enhances B cell Ia expression might also increase the efficiency with which the antigen-specific helper signal could be delivered, thus increasing the antihapten PFC response.

The nature of the factors present in the interleukin preparations which are responsible for the induction of increased B cell Ia expression have not been conclusively identified. FS6 Con A SN is known to contain at least two factors, BCGF and IL-2. Molecular weight fractionation of FS6 Con A SN demonstrated that there was a good correlation between the B cell Ia inducing activity and BCGF (antiimmunoglobulin costimulator assay) and no correlation with IL-2.[32] This result was consistent with the failure of cloned human IL-2 to induce increased Ia expression at concentrations where it could be shown to enhance the generation of primary in vitro PFC response to sheep erythrocytes. Similarly, cloned mouse IFN$_\gamma$ was also without effect on the induction of increased B cell Ia expression, also at concentrations where it could be shown to enhance the generation of anti-SRBC PFC responses.[32] This result was particularly interesting in light of the ability of cloned IFN$_\gamma$ to induce increased Ia expression and antigen-presenting ability of MØ.[34,35] Our results suggest a parallel system exists for B cells which does not, however, include IFN$_\gamma$.

Two other pieces of evidence support the hypothesis that BCGF is responsible for the induction of increased B cell Ia expression. First, analysis of the Con A SN of a panel of T cell clones and hybridomas has demonstrated that their levels of B cell Ia-inducing activity are closely correlated to the levels of BCGF activity they contain as determined in an antiimmunoglobulin costimulator assay. Second, molecular weight fractionation of the Con A-induced culture supernatant of one of the T cell clones again demonstrates that the B cell Ia inducing activity and BCGF activity coelute in the same column fractions, corresponding to a molecular weight of approximately 20,000.

The nature of the factor present in P388 SN which is responsible for induction of increased B cell Ia expression remains unknown. Sephadex® G75 molecular weight fractionation of ammonium sulfate-concentrated P388 SN reveals a small peak of activity corresponding to a molecular weight of 15,000, however, the bulk of the activity elutes in the column void volume, corresponding to a molecular weight of greater than 80,000. Ia-inducing activities have previously been associated with MØ products, including the recent demonstration of a high molecular weight factor produced by P388D$_1$.[36,37] Characterization of the MØ factor active in our assays awaits further study.

REFERENCES

1. Hoffeld, J. T., Marrack, P., and Kappler, J. W., Antigen-specific and nonspecific mediators of T cell/B cell cooperation. IV. Development of a model system demonstrating responsiveness of two T cell functions to HGG *in vitro*, *J. Immunol.*, 117, 1953, 1976.
2. Marrack, P. C. and Kappler, J. W., Antigen-specific and nonspecific mediators of T cell/B cell cooperation. I. Evidence for their production by different T cells, *J. Immunol.*, 114, 1116, 1975.
3. Marrack, P. C. and Kappler, J. W., Antigen-specific and non-specific mediators of T cell/B cell cooperation. II. Two helper T cells distinguished by their antigen sensitivities, *J. Immunol.*, 116, 1373, 1977.

4. Keller, D. M., Swierkosz, J. W., Marrack, P., and Kappler, J. W., Two types of functionally distinct, synergizing helper T cells, *J. Immunol.*, 124, 1350, 1980.
5. Roehm, N. W., Marrack, P., and Kappler, J. W., Antigen-specific, H-2-restricted helper T cell hybridomas, *J. Exp. Med.*, 156, 191, 1982.
6. Harwell, L., Kappler, J., and Marrack, P., Antigen-specific and nonspecific mediators of T cell/B cell cooperation. III. Characterization of the nonspecific mediator(s) from different sources, *J. Immunol.*, 116, 1379, 1976.
7. Hoffeld, J. T., Marrack, P. C., and Kappler, J. W., Antigen-specific and nonspecific mediators of T cell/B cell cooperation. VI. Spectra of cross-reactivity of two HGG-primed helper T cell subpopulations of anti-HGG plaque-forming cells, *Cell. Immunol.*, 33, 20, 1977.
8. Marrack, P., Swierkosz, J. E., and Kappler, J. W., The role of antigen-presenting cells in effector helper T cell action, in *Macrophage Regulation of Immunity*, Unanue, E. and Rosenthal, A., Eds., Academic Press, New York, 1979, 123.
9. Keller, D. M., Swierkosz, J. E., Marrack, P., and Kappler, J. W., Two T cell signals are required for the B cell response to protein-bound antigens, in *Proc. ICN-UCLA Symp. on T and B Lymphocytes: Recognition and Function*, Academic Press, New York, 1979, 107.
10. Marrack, P., Harwell, L., Kappler, J. W., Kawahara, D., Keller, D., and Swierkosz, J., Helper T cell interactions with B cells and macrophages, in *Recent Developments in Immunological Tolerance and Macrophage Function*, Baram, B., Pierce, C. W., and Battisto, J. R., Eds., Elsevier/North-Holland, New York, 1979, 31.
11. Swierkosz, J., Marrack, P., and Kappler, J. W., Functional analysis of T cells expressing Ia antigens. I. Demonstration of helper T cell heterogeneity, *J. Exp. Med.*, 150, 1293, 1979.
12. Harwell, L., Skidmore, B., Marrack, P., and Kappler, J., Concanavalin A-inducible, interleukin-2-producing T cell hybridoma, *J. Exp. Med.*, 152, 893, 1980.
13. Leibson, H. J., Marrack, P., and Kappler, J. W., B cell helper factors. I. Requirement for both interleukin-2 and another 40,000 mol wt factor, *J. Exp. Med.*, 154, 1681, 1981.
14. Leibson, J. G., Marrack, P., and Kappler, J. W., B cell helper factors. II. Synergy among three helper factors in the response of T cell and macrophage depleted B cells, *J. Exp. Med.*, 129, 1398, 1982.
15. Robb, R. J., Munck, A., and Smith, K. A., T cell growth factor receptors. Quantitation, specificity and biological relevance, *J. Exp. Med.*, 154, 1455, 1981.
16. Malek, T. R., Robb, R. J., and Shevach, E. M., Identification and initial characterization of a rat monoclonal antibody reactive with murine interleukin 2 receptor-ligand complex, *Proc. Natl. Acad. Sci. U.S.A.*, 80, 5694, 1983.
17. Endres, R. O., Kushnir, E., Kappler, J. W., Marrack, P., and Kinsky, S. C., A requirement for nonspecific T cell factors in antibody responses to "T cell independent" antigens, *J. Immunol.*, 130, 781, 1983.
18. Lachman, L., Hacker, M., Blyden, G., and Handschumacher, R., Preparation of lymphocyte-activating factor from continuous murine macrophage cell lines, *Cell. Immunol.*, 34, 416, 1977.
19. Mizel, S., Oppenheim, J., and Rosenstreich, D., Characterization of lymphocyte-activating factor (LAF) produced by the macrophage cell line, P388D1. I. Enhancement of LAF production by activated T lymphocytes, *J. Immunol.*, 120, 1497, 1978.
20. Koopman, W., Farrar, J., and Fuller-Bonar, J., Evidence for the identification of lymphocyte-activating factor as the adherent cell-derived mediator responsible for enhanced antibody synthesis by nude mouse spleen cells, *Cell. Immunol.*, 35, 92, 1978.
21. Hoffmann, M., Macrophages and T cells control distinct phases of B cell differentiation in the humoral immune response *in vitro*, *J. Immunol.*, 125, 2076, 1980.
22. Schimpl, A. and Wecker, E., Replacement of a T cell function by a T cell product, *Nature (London)*, 237, 15, 1972.
23. Zlotnik, A., Roberts, W. K., Vasil, A., Blumenthal, E., LaRose, F., Leibson, H. J., Endres, R. O., Graham, S. D., Jr., White, J., Hill, J., Henson, P., Klein, J. R., Bevan, M. J., Marrack, P., and Kappler, J. W., Coordinate production by a T cell hybridoma of gamma interferon and three other lymphokine activities. Multiple activities of a single lymphokine? *J. Immunol.*, 131, 794, 1983.
24. Kappler, J. W., Skidmore, B., White, J., and Marrack, P., Antigen-inducible, H-2-restricted interleukin-2-producing T cell hybridomas, *J. Exp. Med.*, 153, 1198, 1981.
25. Kappler, J., Kubo, R., Haskins, K., White, J., and Marrack, P., The mouse T cell receptor: comparison of MHC restricted receptors on two T cell hybridomas, *Cell*, 34, 727, 1983.
26. Mitchison, N. A., The carrier effect in the secondary response to hapten-protein conjugates. II. Cellular cooperation, *Eur. J. Immunol.*, 1, 18, 1971.
27. Roehm, N. W., Marrack, P., and Kappler, J. W., Helper signals in the plaque forming cell response to protein-bound haptens, *J. Exp. Med.*, 158, 317, 1983.
28. Parker, D. C., Induction and suppression of polyclonal antibody responses by anti-Ig reagents and antigen nonspecific helper factors, *Immunol. Rev.*, 52, 115, 1980.

29. Howard, M., Kessler, S., Chused, T., and Paul, W. E., Long-term culture of normal mouse B lymphocytes, *Proc. Natl. Acad. Sci. U.S.A.*, 78, 5788, 1981.
30. Kappler, J. W., Leibson, J., Roehm, N., Zlotnik, A., Gefter, M., and Marrack, P., Multiple helper T cell activities in B cell responses, in *Progress in Immunology V*, Yamamura, Y. and Tada, T., Eds., Tokyo, 1983, 683.
31. Leibson, H. J., Gefter, M., Ziotnik, A., Marrack, P., and Kappler, J. W., Role of γ-interferon in antibody producing responses, *Nature (London)*, 309, 799, 1984.
32. Roehm, N. W., Leibson, H. J., Zlotnik, A., Kappler, J., Marrack, P., and Cambier, J. C., Interleukin induced increase in Ia expression by normal mouse B cells, *J. Exp. Med.*, 160, 679, 1984.
33. Shimonkevitz, R., Kappler, J., Marrack, P., and Grey, H., Antigen recognition by H-2-restricted T cells. I. Cell-free antigen processing, *J. Exp. Med.*, 158, 303, 19.
34. Zlotnik, A., Shimonkevitz, R. P., Gefter, M., Kappler, J., and Marrack, P., Characterization of the γ interferon mediated induction of antigen presenting ability in P388D₁ cells, *J. Immunol.*, 131, 2814, 1983.
35. King, D. P. and Jones, P. P., Induction of Ia and H-2 antigens on a macrophage cell line by immune interferon, *J. Immunol.*, 131, 135, 1983.
36. Hoffman, M. K., Koenig, S., Mittler, R. S., Oettgen, H. F., Ralph, P., Galanos, C., and Hammerling, U., Macrophage factor controlling differentiation of B cells, *J. Immunol.*, 122, 497, 1979.
37. Walker, E., Maino, V., Sanchez-Lanier, M., Warner, N., and Stewart, C., Murine gamma interferon activates the release of macrophage derived Ia inducing factor which transfers Ia inductive capacity, *J. Exp. Med.*, 159, 1532, 1984.

Chapter 5

GROWTH AND DIFFERENTIATION OF B CELL CLONES IN VITRO

Beverley L. Pike and G. J. V. Nossal

TABLE OF CONTENTS*

---

* Abbreviations used: AECM-Ficoll, aminoethylcarbamylmethylated-Ficoll; AFC, antibody forming cell;
BA, *Brucella abortus*; B cell, bone marrow-derived (B) lymphocyte; BGDF, B cell growth and differentiation factor(s); CAS, Con A-stimulated spleen CM; cIgM, cytoplasmic IgM; Con A, Concanavalin A;
DXS, dextran sulfate; EL-BGDF or EL-BGDF-pik, BGDF contained within EL4-CM; EL4-CM, mediated conditioned by Con A-stimulated EL4 cells; FACS, fluorescence-activated cell sorter; FLU, fluorescein; Ig, immunoglobulin; IL-2, interleukin 2; ISC, immunoglobulin secreting cell; KLH, keyhole
limpet hemocyanin; LPS, *Escherichia coli* lipopolysaccharide; MHC, major histocompatibility complex;
mIg, surface membrane immunoglobulin; PFC, hemolytic plaque-forming cell; POL, polymerized flagellin; T cell, thymus-derived (T) lymphocyte; TD, T cell-dependent; TI, T cell-independent.

I. INTRODUCTION

Differentiation of B cells can be divided into two distinct phases or pathways. The first pathway is an antigen-independent differentiation process, starting with the pluripotential stem cells and proceeding through a series of presently unknown stages to generate a cell which can be identified as being unequivocally committed to the B cell lineage by its expression of cytoplasmic immunoglobulin (cIgM). From this cycling "pre-B" cell, nondividing small lymphocytic cells arise which then acquire surface membrane immunoglobulin (mIg). The literature pertaining to this phase of B cell differentiation has been well reviewed recently by Osmond,[1] Kincade,[2] and Scher.[3] The second pathway of B lymphocyte differentiation is directly relevant to this volume. This is an antigen-driven process, in which virgin, noncycling B lymphocytes are sent along a differentiation pathway by appropriate interaction with specific antigen ultimately to form antibody-secreting plasma cells and/or memory cells.

In our laboratory we have been seeking to enlarge our understanding of the interactions that can take place following the combination of specific antigen molecules with the mIg receptor on the noncycling, virgin B lymphocyte. Such interactions can be activating, in which case the lymphocyte concerned undergoes blast cell transformation, repeated division, and differentiation toward a cell specialized for antibody production. Alternatively, the interaction may lead the lymphocyte concerned to register and store a negative signal, lowering its capacity to respond to later activating signals. A great fascination of the immune system is that it consists of single elements, the clonally individuated lymphocytic cells, each capable of a unique act of recognition. It also consists of a precisely regulated network of cellular interactions exerting intricate control in both a positive and negative manner. We have actively sought to develop a strategy that is inherently quite simple. Murine lymphoid cell populations are fractionated by an antigen-affinity procedure[4] to yield a subpopulation of hapten-specific B cells, which, if desired, can be further subfractionated on the basis of avidity for antigen or other characteristics, e.g., antigen avidity,[5] size,[6] or mIgD status.[7] Either the donor's age or the organ source can be varied to yield B cells of defined maturity. We have then examined the behavior of these single hapten-specific B cells on interaction with antigen at the clonal level using the variety of cloning systems available, some in which the single B cell is stimulated in total isolation, and others in which T cells are intentionally added. By this approach, which is still in its early stages, we hope that we may eventually be able to identify many of the single elements which exert an influence on the process of antigen-driven B cell differentiation at the level of the single antigen-specific B cell.

In this chapter we review the pertinent recent findings relating to the factors controlling the growth and differentiation of the single B cell grown as a clone in vitro. Much of the discussion will center around our own findings on the effects of antigenic stimuli, lymphokines (and other cytokines), T cells, and other accessory cells on the activation and clonal differentiation of single hapten-specific B cells. Before dealing with the recent knowledge generated in single cell cloning systems, we will briefly review the developments that led to the generation of those systems.

A. Factors Influencing B Cell Differentiation In Vitro

Many of the factors which may influence whether a particular B cell will be activated into the differentiation pathway on interaction with specific antigen have been well identified in the literature. One of these is the heterogeneity within the B cell population per se. B cells are heterogeneous with respect to their differentiation or maturational status;[5,7,8] cycling status, size, or density;[5,10-19] whether they belong to a particular lineage, for example, Lyb 5;[12,20-22] or possess mIgD;[9,23-28] and the possibility of belonging to distinct subsets responding to either T cell-dependent (TD) or T cell-independent (TI) antigenic stimuli.[7,18,29-33] Some of these issues will be addressed later in this chapter.

In recent years most attempts to gain detailed insight into the mechanisms of B lymphocyte activation have been via the use of complex in vitro bioassay systems in which lymphocytes are triggered into blastogenesis and clonal proliferation either by antigen (exerting its effects on cells with receptors for that antigen) or by mitogens (exerting their effects on particular cellular subsets regardless of receptor idiotype).

Much of the work has been centered on the respective roles of antigen or mitogen, on the one hand, and antigen-specific or nonspecific T cell-derived or macrophage-derived regulatory factors,[34-46] on the other, in the cascade of cellular events involved in an immune response. It is evident that antigen, acting alone on a B cell with surface immunoglobulin (mIg) receptors reactive with it, cannot promote all the necessary elements of the cascade[45-49] and that additional factors are required. It has been proposed that at least two distinct sets of factors govern the processes of proliferation and of differentiation. This concept will be discussed later.

Antigens have been frequently grouped into "T cell-independent" (TI) or "T cell-dependent" (TD) on the basis of their alleged requirement for the collaborative help of major histocompatibility complex (MHC) restricted antigen-specific T cells to initiate immune induction.[12,50-52] In the standard dense Mishell-Dutton[53] type tissue cultures, the effects of triggering stimuli are frequently read out by counting the number of antibody-forming cells at the end of a 3- to 5-day culture period. However, such cultures are influenced by a variety of intercellular interactions and feedbacks, both positive and negative, and so several groups including our own have placed increasing emphasis on B cell cloning techniques.

B. Systems for Clonal Analysis of B Cell Activation

In vitro, limiting dilution cloning methods, where precursor B cells are isolated from many of the effects of the immune network, has provided a most useful approach to the assessment of B cell competence and activation leading to blastogenesis, clonal expansion, and differentiation. Well-defined cloning methods exist for both TI and TD stimulation. Following some preliminary work by Marbrook,[54] B cell cloning methods for T cell-independent triggering were defined in our laboratory. In these systems,

B cell-depleted thymus cells were used as a "filler" cell population in 0.2-mℓ microculture wells.[55-57] This approach has also been widely used by other groups for clonal analysis of TI B cell activation.[24,58-62] Furthermore, mitogen-driven B cell proliferation can be assessed clonally in agar gels, where proliferating B cells form loose colonies that can be enumerated.[63-68] The cloning efficiency of this system is raised by the addition of macrophages.[66,68]

Two main cloning methods have been used for the study of TD triggering. One widely used method is the splenic microfocus assay of Klinman.[69,70] Responder B cells are adoptively transferred into carrier-primed, lethally irradiated mice and 1 day later the host spleens are removed and cut into approximately 50 fragments which are cultured with appropriate antigens, e.g., hapten-carrier complexes. When limited numbers of B cells are used, it becomes a clonal assay. Antibody formation is detected by analysis of culture supernatants for specific antibody using a sensitive radioimmunoassay (RIA) procedure. This approach has allowed both accurate enumeration of frequencies of antigen-specific antibody-forming cell (AFC) clone precursors within given B cell populations and insight into the clonal product, such as isotype class and kinetics of class switching, etc.[16,17,70-76] Other widely used TD cloning methods are based on the method of Waldman, Lefkovits and Quintans.[77,78] In this liquid microculture method, limiting numbers of B cells interact with nonlimiting numbers of carrier-primed T cells. Both these TD systems have been further refined to use not T cells from living animals to provide specific T cell help, but T cells from antigen-specific in vitro cloned helper T cell lines to deliver the MHC-restricted signal.[79,80]

All these cloning systems have a major disadvantage in that they largely depend upon the presence of thymic cells, macrophages, or X-irradiated splenic cells for the support of clonal proliferation and differentiation, each population in itself capable of exerting its own unknown effects on the clonal growth.[34-46,66,68] Furthermore, they allow only assessment of the frequency of precursor cells inducible to differentiate to AFC status and do not allow assessment of the intermediate differentiative stages of blastogenesis and proliferation. Despite these limitations, the methods have provided a useful approach to enumerating the frequency of B cells inducible with specific stimuli and assessment of the clonal product in both a quantitative and qualitative manner.

II. A MITOGEN-DRIVEN SYSTEM FOR CLONAL GROWTH OF SINGLE B CELLS IN VITRO

Recently, a major advance was achieved in B lymphocyte cloning technology which provided the opportunity for a new approach to dissecting some of the mechanisms influencing the process of immune induction in B cells. Kettman and Wetzel[81,82] described conditions under which single, isolated murine splenic B cells could be stimulated to proliferate to form clones in the absence of any other cell types. Stimulation with a mixture of the mitogens *Escherichia coli* lipopolysaccharide (LPS) and dextran sulfate (DXS) in 10 μℓ Terasaki culture wells was reported to induce up to 70% of single B cells to proliferate.[82] Proliferation was easily monitored by microscopic examination as the absence of filler cells allowed visualization. Assay of culture supernatants, using a hemolysis in gel spot test with protein A-coupled sheep erythrocytes,[83] showed only a few clones to be secreting antibody. Proliferation was shown to be dependent upon the batch of fetal calf serum used;[81,84] deficient batches could be made supportive by the addition of supernatant from a monoclonal Dennert T cell line C.C3.11.75[84] known to contain both B cell growth-promoting[85] and differentiation-promoting[43,86] factors. The addition of peritoneal exudate cells or splenic-derived adherent cells also raised the cloning efficiency[81] and, in some instances, resulted in increased numbers of AFC clones.

Our own experience with LPS + DXS stimulation of unfractionated splenic cells essentially confirmed Kettman and Wetzel's findings with a few modifications. Somewhat lower cloning efficiencies were obtained in a series of 5 experiments; $15.2 \pm 1.8\%$ of all spleen cells (i.e., 30% of B cells) formed proliferating clones as opposed to claims of 27 to 70% of B cells.[82] Furthermore, antibody formation was more frequently observed, with approximately half of the clones generated shown to contain immunoglobulin secreting cells (ISC) as detected by the protein A reverse plaque assay.[83] No dependence upon a particular fetal calf serum batch was noted. Our experience with single hapten-specific B cells in this proliferation-geared, mitogen-driven system will be dealt with later in this chapter.

This development by Kettman and Wetzel provided for the first time a system with the potential of uncoupling the events of blastogenesis, proliferation, and differentiation in a simple and potentially quantitative manner at the level of the single precursor B cell. It also demonstrated that under the inductive conditions used, the process of proliferation and differentiation could be considered distinct entities.

III. AN ANTIGEN-DRIVEN SYSTEM FOR GROWTH AND DIFFERENTIATION OF SINGLE HAPTEN-SPECIFIC B CELLS IN VITRO

A. Experimental Approaches

Since 1975, we have sought to understand the biology of B lymphocyte activation using virgin B cells selected for their reactivity to a particular hapten using the simple hapten-gelatin fractionation procedure described by Haas and Layton[4] in our laboratory. Normal murine spleen cells are vigorously rocked for 15 min at 4° in petri dishes coated with a thin layer of hapten-gelatin, the nonadherent cells removed by vigorous washing and the adherent cells recovered by melting the gelatin at 37°. Collagenase treatment is used to remove adherent antigen from the surface of the cells. This procedure, which has been described in detail elsewhere,[5,6,56,57] selects a population of B cells representative of the total B cell population with respect to size distribution and mIg avidity profile (Figure 1). The population is 97% B cells,[5,68,87] and with fluorescein (FLU) acting as the hapten, 60 to 80% of the binding population has been shown to be truly FLU-specific[6,87] and to exhibit a wide range of FLU-binding avidities.[5,6,87] Hapten-gelatin binding B cell populations have been extensively shown to be highly enriched for functional activity.[5-7,40,45,56,57] The obvious development of the Kettman-Wetzel single cell, mitogen-driven system was into an antigen-driven system in which single hapten-specific splenic B cells could be activated by specific antigen to proliferate and differentiate to antibody-forming status. Such a system would allow dissection of the influence of defined stimuli on separate events such as blast cell transformation, progressive clonal expansion, and differentiation to AFC status, unimpeded by the unknown effects of the stimuli on other non-B cells in the culture.

The culture procedures used in the original work to be presented have been described in detail elsewhere.[6,40,45,87] Single B cells frequently grew as clusters, but sometimes as dispersed clones or a mixture of both. A well was scored positive for proliferation if one or more clusters of more than three blast cells were observed, or if the number of cells present was clearly greater than the input number. With stronger stimuli, many clones contained 100 or more cells.

A total in situ approach to the assay of AFC, as well as proliferating clones, was developed for the purposes of filler cell-free systems allowing large numbers of replicate cultures to be screened.[89] The procedure, as shown diagrammatically in Figure 2, allowed the various parameters influencing clonal growth and differentiation to be assessed by limiting diluting analysis.

Assay for anti-FLU AFC clones was as previously described.[40,45,60,88]

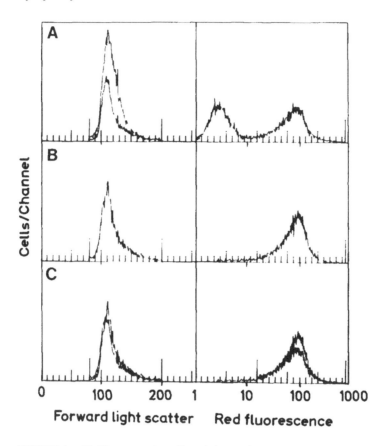

FIGURE 1.   FACS generated profiles of nonfractionated spleen cells (A) and FLU-gelatin binding spleen cells (B) from adult donors labeled with rhodaminated anti-Ig. The parameters shown are forward light scatter (size) on a linear scale (left-hand panels) and red fluorescence (mIg) on a logarithmic scale (right-hand panels). The y axis for all panels is cells/channel presented on a linear scale. The forward light-scattering profile of all splenic mIg⁺ cells is also shown in panel A and again in panel C superimposed on the forward light scattering of FLU-gelatin binding B cells (from panel B). Nonviable cells were excluded by PI positivity.

B. Requirements for Antigen-Driven B Cell Growth and Differentiation

1. Development of System

The antigen-driven filler-cell free system was defined in the first instance using the TI antigen, FLU-polymerized flagellin (FLU-POL), as the antigenic stimulus.[40] FLU-POL was virtually nonstimulatory when acting alone on FLU-gelatin-enriched B cells. The further addition of a crude mixture of factors contained within Concanavalin A (Con A)-stimulated spleen cell (CAS) conditioned medium (CM) resulted in a significant proportion of B cells proliferating to form clones, the majority of which contained anti-FLU IgM AFC. CAS alone was virtually inactive. The anti-FLU AFC cloning efficiency was similar to that previously achieved with the standard thymus-"filler" cell-supported cloning system,[55-57] being around 3 to 6%. Evidence that the target of action was the hapten-specific B cell was shown by linear regression analysis, which showed only a single entity to be limiting.

This antigen-driven system permits various cytokines to be assessed for the presence of B cell-active factors, with a single B cell as the unequivocal target cell. For example, FLU-POL acting alone is virtually ineffective; FLU-POL acting in synergy with B cell

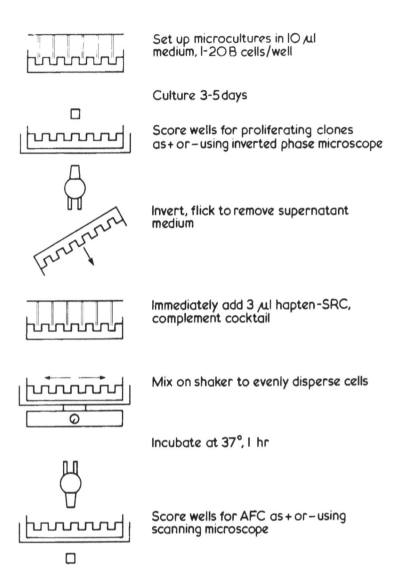

Set up microcultures in 10 µl
medium, 1-20 B cells/well

Culture 3-5 days

Score wells for proliferating clones
as + or – using inverted phase microscope

Invert, flick to remove supernatant
medium

Immediately add 3 µl hapten-SRC,
complement cocktail

Mix on shaker to evenly disperse cells

Incubate at 37°, 1 hr

Score wells for AFC as + or – using
scanning microscope

FIGURE 2.   Schematic representation of *"in situ"* assay method used for detection
of proliferating and PFC clones in 10 µl well filler cell-free microcultures.

growth and differentiation promoting factors is able to promote significant clonal expansion and differentiation to AFC status.

*2. B Cell Growth and Differentiation Factors (BGDF) Are Distinct from Interleukin-2
(IL-2)*

CM prepared by Con A stimulation of certain cloned T cell tumor lines or hybridomas was screened for its ability to produce factors capable of promoting FLU-POL-driven growth and/or differentiation of isolated FLU-specific B cells. CM prepared from the thymoma cell line, EL4, was found to be active in the system.[45] The frequency of proliferating clones with FLU-POL alone of 0.5% (95% confidence limits 0.3 to 0.7%) was increased 13-fold to 6.3% (4.9 to 8.0%) by the addition of EL4-CM. With medium alone or with EL4-CM alone, only 0.1 to 0.2% of wells contained clones. Most clones contained anti-FLU AFC.

On purely operational grounds, this synergistic capability was termed "B cell growth and differentiation factor(s)'', or BGDF.[45] This bioactivity has since been termed EL-BGDF-pik.[89,90] EL4-CM from unstimulated EL4 cells contained no significant BGDF activity. Biochemical studies showed the EL4-BGDF activity to elute over a broad range on gel filtration (30 to 60,000 daltons), to be unaffected by treatment with 6 M guanidine HCl, and to be distinct from interleukin 2 (IL-2).[45]

To avoid any ambiguity in reference to the target of action of antigen and EL-BGDF-pik, cultures containing an average of 1.5 to 2 cells per 10-μl well were set up, and 18 hr later wells containing exactly 1 cell were noted. This approach circumvented the need for Poissonian statistics. Most proliferating clones contained anti-FLU AFC (Table 1). Both FLU-POL and EL-BGDF-pik were essential for the induction and maintenance of both proliferation and differentiation. These results, in fact, provided for the first time unequivocal evidence that a T cell-derived lymphokine could synergize with specific antigen to promote the in vitro growth and differentiation of a B cell.

BGDF activity was also detected in some T cell hybridomas.[45] CM from a T cell hybridoma, which produces a granulocyte/macrophage colony-stimulating factor but not IL-2, and a myelomonocytic leukemia cell line, WEHI-3, that produces factor(s) active on a range of hemopoietic progenitor and stem cells and a factor resembling lymphocyte-activating factor or interleukin 1 (IL-1) but not IL-2, were both found to be negative in this system.[45]

C. Comparison of Antigen-Driven and Mitogen-Driven Clonal Growth and Differentiation

Figure 3 shows a typical experiment with FLU-POL + BGDF stimulation in which 7.2% (95% confidence limits 5.5 to 8.9%) of FLU-gelatin fractionated cells proliferated and 3% (2.1 to 4%) formed clones containing anti-FLU AFC. In contrast, the frequency of anti-FLU AFC clone precursors among the nonfractionated cells was 228 × 10^{-6} (175 to 281 × 10^{-6}). Proliferation could not be assessed as irrelevant cells obscured visualization.

In 26 separate experiments of this design, the FLU-gelatin fractionated (FLU-enriched) cells yielded a mean frequency value of 5.2 ± 0.5% for proliferating clones and 3.5 ± 0.4% for anti-FLU AFC clones. In 4 separate experiments, nonfractionated spleen cells yielded 187 ± 26 × 10^{-6} anti-FLU AFC clones. The mean functional enrichment factor provided by FLU-gelatin fractionation in the EL-BGDF-pik supported system is 187-fold, consistent with prior experience with thymus-filler cells.[55-57]

In a series of 21 separate experiments, FLU-enriched splenic B cells were stimulated with LPS + DXS; 25.2 ± 2.8% of cells proliferated and 4.8 ± 0.5% formed clones containing anti-FLU AFC. Figure 4 shows an experiment in which FLU-POL + BGDF and LPS + DXS were directly compared. Thus, although LPS + DXS induced significantly more proliferation than FLU-POL + BGDF, a similar number of clones containing anti-FLU AFC were generated.

Proliferation induced among nonfractionated spleen cells by LPS + DXS showed 15.2 ± 1.8% (5 experiments) of all spleen cells to form clones. Adjusted for the B cell content of the input population, this represents a similar frequency to that obtained using the FLU-gelatin-enriched population. Assay for anti-FLU AFC clones yielded a frequency value of 327 ± 65 × 10^{-6}, this a slightly higher value than obtained with antigen, possibly being due to induction of lower affinity splenic FLU-specific B cells by the mitogenic stimulus.

Studies performed to establish whether either system was more geared to proliferation or to antibody formation showed that although a higher frequency of FLU-enriched B cells proliferated with LPS + DXS, generating three times more B cells than FLU-POL + BGDF stimulation, there was little difference in the number of cells per clone in either system.[6] The majority of cells generated were shown not be AFC as

Table 1
PROLIFERATION AND GENERATION OF ANTI-FLU AFC CLONES BY SINGLE, ISOLATED FLU-ENRICHED B CELLS STIMULATED WITH FLU-POL + EL4-BGDF[a]

Number of wells	600
Mean input number of cells per well	1.8
% Cell death first 18 hr	34
Wells with exactly one cell at 18 hr	217
Above wells showing proliferation[b]	20 (9.2%)
Above wells showing AFC formation[b]	18 (8.3%)

[a] A simultaneous Poissonian study on the same cell population cultured at 20 per well yielded the following background frequencies, for % proliferation and % anti-FLU AFC clones, respectively: medium alone, 0.08, 0.08; EL4-CM, 0.04, 0.17; FLU-POL, 0.76, 0.81.
[b] Only wells containing exactly one cell at 18 hr included.

Reprinted from Vaux, D. L., Pike, B. L., and Nossal, G. J. V., *Proc. Natl. Acad. Sci. U.S.A.*, 78, 7702, 1981. With permission.

assessed using a specific anti-FLU AFC or nonspecific (ISC) plaque assay. In both systems, more cells proliferated than matured to AFC status. The antigen-driven system was slightly biased toward maturation.

D. Effect of the Addition of Thymus Filler Cells on Cloning Efficiency
Thymus cells had been extensively used as "filler" cells for the support of antigen-dependent cloning of antihapten AFC precursors in many of our previous studies.[55-57] The cloning efficiencies achieved were similar to those reported with the filler cell-free antigen-driven plus lymphokine system. Cultures in which single virgin B cells are held in complete isolation could be suboptimal due to deficiencies not only in tissue-specific growth factors, but also in the largely ill-defined, nonspecific growth-promoting influences of the feeder cells which comprise an essential component of many in vitro cloning systems. For this reason, the effect of adding 10^5 thymus cells to these 10-μl single B cell cloning systems was investigated. Mice from the CBA/N strain were chosen as thymus donors, as the CBA/N B cells in our hands are so poorly unresponsive to all stimuli,[91] and thus B cell contamination of the thymus population was an insignificant factor. As the presence of thymus fillers obscured visualization, only antibody formation could be directly compared. These experiments provided surprising results in that an extraordinary synergy was observed between EL4-CM and the thymocytes, resulting in marked enhancement of the antibody response of FLU-specific B cells.[6]

The "filler" effect is not MHC-restricted, as the cloning efficiency of allogeneic and syngeneic B cells are enhanced to an equivalent degree. The FLU-POL-elicited response of CBA FLU-enriched B cells was enhanced from 8.6% (53 PFC per 100 input B cells) to 20.0% (314 PFC per 100 input B cells), and that of BALB/c FLU-enriched B cells from 6.3% (64 PFC per 100 input B cells) to 21.3% (239 PFC per 100 input B cells). The effect was also shown to be not unique to thymus cells from CBA/N donors (data not shown); CBA/N-derived thymus cells were used only because of the extremely low frequency of responsive B cells in the thymus cell population. Similar enhancement was seen with LPS + DXS[6,89] and with other TI antigens.[89,91] One in 700 unfractionated spleen cells formed an anti-FLU AFC clone when stimulated with FLU-POL + EL-BGDF in the presence of thymus fillers.

In collaboration with Dr. J. J. Cebra, the frequency of anti-FLU AFC clones was assessed using an antigen-specific, sensitive RIA procedure which allows the detection of specific antibody in the 10-μl culture supernatant fluid rather than a direct anti-

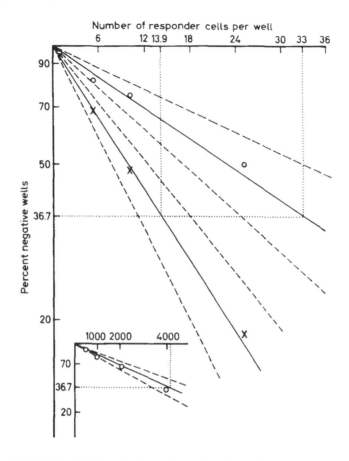

FIGURE 3.   Limiting dilution analysis of cloning efficiency of nonfractionated spleen cells and FLU-gelatin fractionated splenic B cells stimulated by FLU-POL + BGDF in vitro. Proliferating clones (x—x) and anti-
FLU PFC clones (O—O) generated by the FLU-gelatin binding B cells
are shown. The anti-FLU PFC clones generated by nonfractionated
spleen cells are shown in the inset. The dashed line indicates the 95%
confidence limits and the number of cells required per well to yield 36.7%
negative wells is indicated by the dotted line. Frequency values are given
in the text. (Reprinted from Pike, B. L., Vaux, D. L., and Nossal, G. J.
V., *J. Immunol.*, 131, 554, 1983. With permission.)

FLU IgM plaque assay. When stimulated with FLU-POL + EL-BGDF in the presence
of 10^5 CBA/N thymus fillers, $55.3 \pm 5.5\%$ of single FLU-enriched splenic B cells
formed an anti-FLU AFC clone.

The enhancement of TI responses by the addition of thymus cells has been reported
by others.[92] The mechanisms involved are unclear. On the addition of the thymus fillers
to the EL-BGDF + FLU-POL system, it immediately became apparent that a component of the conditioned medium allowed extensive proliferation of some cells within
the thymus filler population which may have produced factors that acted directly on B
cells. This raised the possibility of a lymphokine cascade. Interestingly, however,
cloned human IL-2 does not enhance the FLU-POL + thymus response.[93] Furthermore, enhancement of the FLU-LPS and LPS + DXS antibody response occurred in
the absence of BGDF without any significant proliferation of the thymus cell population, rendering the significance of thymic cell proliferation highly dubious. Macrophages or dendritic cells within the thymus filler population could have possibly exerted growth-promoting effects.

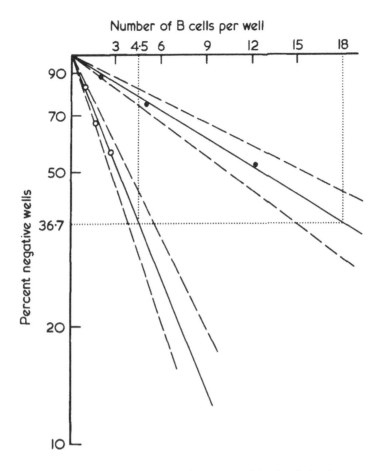

FIGURE 4. Limiting dilution analysis of the proliferation induced among FLU-gelatin fractionated splenic B cells in vitro; FLU-POL + EL-BGDF-pik (•—•) and the mitogens LPS + DXS (O—O). The dashed lines represent the 95% confidence limits. The frequence values were 5.6% (4.5 to 6.7%) for FLU-POL + BGDF and 22.4% (18.2 to 26.5%) for LPS + DXS. Values for anti-FLU AFC clone formation (not shown) were 3.6% (2.8 to 4.4%) for FLU-POL + BGDF and 4.2% (2.5 to 5.8%) for LPS + DXS. (Reprinted from Pike, B. L. and Nossal, G. J. V., *J. Immunol.*, 132, 1687, 1984. With permission.)

Finally, nonimmunological phenomena such as cell crowding or other nonspecific influences could be responsible. Whatever the mechanism, there was marked synergy with all effective forms of stimuli. In fact, both the proportion of cells developing into AFC clones and the number of AFC per clone were by far the highest yet encountered after many years' work with hapten-gelatin fractionated cells.[6,89]

The addition of filler cells certainly obscures identification of the target of action of exogenously added factors. Enhancement of the FLU-POL-driven response has been achieved using CM which are inactive when acting with FLU-POL on isolated single FLU-specific B cells. These results are mentioned more as a cautionary tale, because factors having no *direct* effects on single isolated B cells now could be shown to have significant effects, presumably by initiating the lymphokine cascade,[44] i.e., activating cells within the complex population of thymus filler cells (T cells, macrophages, accessory cells, etc.) to produce substances active on B cells. The remarkably high degree of antibody formation from single B cells under these conditions at least offers a maximal target in the reconstruction of the single elements involved.

IV. CLONAL GROWTH AND DIFFERENTIATION OF B LYMPHOCYTES OF VARIOUS SIZES

Much interesting work has recently been done on the influence of cell cycle stage on B cell triggering.[10,11,13-15] Emphasis has been placed on the differing requirements for activation of small, resting B cells and of large, preactivated B cell blasts, using both TI and TD nonclonal in vitro assay systems.[10,13,49,95] In general, these two populations have been regarded as distinct, often being termed small or large B cells. Out of this, a tempting consensus might be reached that small, resting, G_0 B cells can be activated in one of two ways, namely, by a strong mitogenic stimulus such as LPS,[11] or by appropriate TD antigen plus T cell help.[12,13,96] The bulk of T-independent cells would be slightly larger cells, not in S phase, but possessing increased RNA and in the G_1 phase of the cycle. Some of the differences between the requirements for activation of "small" vs. "large" B cells, in studies where the latter in fact are represented by large, mitogen-preactivated B cell blasts, could be explained by the expression of receptors for BGDF on the surface of the activated "cycling" B cells, allowing them to more readily progress along the proliferative/differentiation pathway.

To this end, we have done some preliminary experiments in which cells were subjected to 24 hr treatment with LPS + DXS, washed, and recultured singly either with a continuation of LPS + DXS or else with factors in the absence of antigen. In essence, we posed the question as to whether EL-BGDF-pik could support the continued proliferation and further differentiation of cells that have already been thrown into active mitotic cycle and have, presumably, expressed receptors for growth and differentiation factors. The results (Table 2) show that residual mitogen, presumably adherent to the cells, can lead to continued proliferation and differentiation when acting alone. The additional presence of EL-BGDF significantly augments the continued proliferation and differentiation.

A. Clone Development by Large and Small Hapten-Specific B Cells

The availability of hapten-specific target virgin splenic B cells, which could be further fractionated into cohorts of various sizes on the basis of their forward light scattering properties using the FACS, prompted us to examine the capacity of B cells of various sizes to perform in the various single-cell cloning systems.[6] As expected on the basis of the literature, the small (G_0) cells performed poorly when stimulated with FLU-POL + BGDF, whereas the cells with larger than median size (G_0* or G_1 cells) performed significantly better, contributing to greater than 90% of the total response both in terms of proliferative and antibody-forming capacity. More surprisingly, when LPS + DXS were used as stimuli, through the small cells performed relatively better, the larger cells still contributed 76% of the proliferating and 86% of the antibody-forming response. From 42 to 59% of the larger subset formed proliferating clones compared to 7 to 17% of the smaller subset. The larger cell subset required a specific stimulus, either FLU-POL + EL-BGDF or LPS + DXS for initiation into clonal growth. As distinct from preactivated blast cells (Table 2), EL-BGDF-pik alone was virtually nonstimulatory when acting on the G_0 or G_1 cell population. This challenges the current paradigm that such cells, having been activated in some way in vivo, now require only lymphokine factors to complete their clonal expansion and differentiation.

1. Cloning Efficiency in Antigen-Driven Thymus Filler System

The effect of thymus filler cells on the activation of the various size subsets was investigated, mainly to see if their presence could enhance the response elicited from the smaller cells. As discussed above, addition of thymocytes markedly enhances the

Table 2
EFFECT OF BGDF ON FLU-ENRICHED B CELL
BLASTS[a]

Stimulus	% B cell blasts proliferating	% Anti-FLU AFC clones
Nil	12.8 ± 3.4	4.6 ± 1.6
Nil + LPS + DXS	41.9 ± 11.4	8.9 ± 2.8
EL-BGDF-pik	17.3 ± 5.7	15.9 ± 4.8

[a] FLU-enriched B cells were stimulated for 24 hr with LPS + DXS, then recultured after extensive washing with medium alone, LPS + DXS, or medium containing 5% EL-BGDF-pik.

antibody response of FLU-POL-stimulated, single FLU-specific B cells when EL-BGDF-pik is also present.[6] Table 3 shows the antibody response of FLU-specific B cells of various sizes when stimulated with FLU-POL + EL-BGDF-pik, alone or in the presence of 10^5 CBA/N-derived thymus filler cells. A five- to sixfold increase in clone frequency was achieved by the addition of the thymus cells, irrespective of the size of the responding FLU-specific B cell. The larger cell subset still contributed 78% of the total response. As shown in Table 3 and as reported by us elsewhere,[6,89,91] this thymus filler cell-supported cloning system is capable of yielding from seven to ten antigen-specific AFC per B cell originally placed into culture if hapten-specific cells slightly larger than the median are used.

Analysis of culture supernatant fluid using a RIA procedure has yielded an anti-FLU AFC clone frequency of 55.3 ± 5.5% with nonsize fractionated FLU-enriched B cells. Preliminary results indicate that this median figure reflects an approximate 20% cloning efficiency of the smaller cells and an approximate 80% for the larger cells. These high cloning frequencies raise questions about TI- and TD-responsive B cells as distinct separable subsets either according to their size,[5,10-19] IgD status,[9,23,24] or their Lyb 5 status.[12,20-22] This issue will be addressed later in this chapter.

Studies reported in the literature utilizing the combined approach of cell separation technologies and clonal analysis, and TI, TD antigenic, or mitogenic challenge systems, has shown at least half of the total activity of splenic B cell populations to reside in the medium to large cell subset.[6,16,17] In fact, few, if any, studies have provided evidence for distinct subsets amongst nonprimed B cells, using analysis of responsiveness at a clonal level. Nonclonal studies[12,13,18] have suggested that TD-responsive B cells are smaller than TI ones.

V. LYMPHOKINE DEPENDENCE OF VARIOUS TI ANTIGENS FOR INDUCTION OF DIFFERENTIATION OF SINGLE B CELLS

Following the elucidation of the respective roles of T and B lymphocytes in the antibody response, reviewed by Miller,[97] it was soon realized that some antigens and mitogens could induce antibody formation in T cell-deficient mice[98,99] or in tissue culture.[100] This led to the postulation of two types of antigens, TI and TD, and raised the issue of whether there might be two corresponding lineages of TI- and TD-responsive B cells.[7,18,29-33] TI antigens were originally thought to activate B cells in a manner which by-passed the requirement for T cell help, clearly delineating them from TD antigens which exhibited an absolute requirement for MHC-restricted, antigen-specific T cell help.[12,50-52]

TI antigens have since been further classified into TI-1 and TI-2 antigens based on their ability to elicit antibody responses from normal adult and suckling mice and from

Table 3

CLONING EFFICIENCY OF SIZE-FRACTIONATED FLU-SPECIFIC B CELL
POPULATIONS STIMULATED WITH FLU-POL AND BGDF, IN THE PRESENCE OR
ABSENCE OF THYMUS FILLER CELLS[a]

		Without fillers		With fillers		
FACS fraction[b]	% Of total population	% AFC clones[c]	Contribution to total response	% AFC clones	Contribution to total response	AFC/100 input B cells[d]
Unsorted FLU-specific cells	100	3.0 ± 0.9[e]	100	14.7 ± 2.9	100	178 ± 51
Small	52—80	0.8 ± 0.2	14	6.2 ± 1.2	22	68 ± 16
Large	20—48	8.8 ± 1.4	86	40.7 ± 6.9	78	720 ± 70

[a] Responses generated by FLU-gelatin binding B cells when stimulated with 0.1 μg/mℓ of FLU-POL in the presence of EL-BGDF-pik (5% v/v) in 10 μℓ of medium either alone or in the added presence of 10⁵ CBA/N thymus cells.

[b] FLU-specific B cells were sorted into size cohorts using FACS as described elsewhere.[6] Small cells fell between channels 80 to 110 and large between channels 120 to 200.

[c] % Cells forming anti-FLU-AFC clones as assessed by directly FLU-specific hemolytic plaque assay.

[d] Determined as % clone frequency × average clone size. This figure is an underestimation due to difficulty in dispersing clusters containing AFC.[56]

[e] Values represent mean ± SEM of three to four experiments.

adult CBA/N mice affected by an X-linked immunodeficiency (xid) syndrome which is expressed as a defect in B cell maturation (reviewed by Scher[101]). TI-1 antigens, such as haptenated-lipopolysaccharide (LPS) or haptenated-*Brucella abortus* (BA), are claimed to have the ability to stimulate antibody responses in all three sets of animals, whereas TI-2 antigens, of which haptenated-AECM Ficoll or haptenated-dextran are prototypes, should effectively stimulate only normal adult mice.[8,9,21,102,103] On this basis, TI-2 antigens are seen as capable of stimulating only a more mature type of B cell absent in CBA/N mice and in newborn mice.[8] Haptenated-polymerized flagellin (POL), used extensively in our laboratory, has been considered a TI-2 antigen.[21]

This picture soon became more complex when it was found that lymphokine-rich CM were obligatorily required in some TI responses[104-106] or greatly augmented others.[106-108] Furthermore, a requirement for Ia-positive adherent cells or macrophages has been demonstrated in some systems.[108-110] Recent studies have also shown the responses to some TI antigens to be partially dependent on T cells,[105,111] but in a nonantigen-specific, non-MHC-restricted manner. Thus, the simple picture of TI antigens directly triggering the B cell required modification, and even the names TI-1 and TI-2 were challenged by some of their early proponents.[104] It appeared that small numbers of T cells and/or macrophages contaminating dense tissue cultures had posed real problems for the interpretation of experiments on B cell-enriched cultures.

A. Activation Requirements of B Cell Subsets at the Level of the Single Hapten-Specific B Cell

Two aspects of our strategy toward understanding B cell activation definitively cut through this methodological impasse: first, the use of an antigen-affinity fractionation procedure, which selects B cell populations enriched for antigen reactivity from both mature and immature cell populations,[5,112] providing B cells of varying maturational status as the potential targets for antigenic activation; second, the availability of a cloning system in which just one single B cell would be the unequivocal target of action for added antigens, mitogens, or lymphokines, thus obviating any arguments of possible indirect effects, such as, for example, lymphokine cascades.[44] These two strategic advantages may be critical both for the reclassification of antigens and responsive B cell subsets, and for the definition of B cell-active growth and differentiation factors.

Accordingly, we have begun to survey a variety of TI stimuli for (1) their ability to induce clonal growth and/or maturation of single hapten-specific B cells in the single-cell cloning systems, in both the absence and presence of intentionally added EL-BGDF-pik; and (2) to study at a quantitative, clonal level the comparative performance of normal adult, hapten-specific B cells (which should respond to both TI-1 and TI-2 antigens); and hapten-specific B cells selected from the spleens of normal neonatal donors and from CBA/N xid donors by hapten-gelatin fractionation. We studied four supposedly TI antigens: FLU-LPS and FLU-BA, as examples of putatively TI-1 antigens,[102,103] and FLU-AECM-Ficoll and FLU-POL, as examples of TI-2 antigens.[21,104] The behavior of one TD antigen, FLU-keyhole limpet hemocyamin (FLU-KLH), and the mitogens LPS and DXS, as well as the carriers BA and AECM-Ficoll, were also examined.

1. Activation of Single FLU-Specific B Cells from Normal Adult Donors

In the first instance, single, isolated, FLU-specific murine splenic B lymphocytes from normal adult donors were used as the unequivocal target cell in 10-$\mu\ell$ cultures unsupported by accessory, feeder, or filler cells, and the TI stimuli were used over a wide range of concentrations both in the presence and absence of BGDF. Both clonal proliferation and differentiation to AFC status were assessed.

The major findings of this survey are shown in Table 4, the studies being reported comprehensively elsewhere.[89] Figure 5 shows the responses obtained when stimuli were compared within a single experiment. The results allowed the triggering stimuli to be divided into four distinct categories: (1) FLU-KLH, a TD antigen, failed to stimulate single B cells even in the presence of BGDF, as did unconjugated AECM-Ficoll. Antigens such as FLU-KLH, FLU-HGG, and doubtless many other hapten-protein conjugates fall into this category. Such antigens can negatively signal MHC-restricted, specific B cells if T cell help is withheld;[31,112,113] (2) FLU-POL and one FLU-Ficoll conjugate (FLU$_{53}$-Ficoll) were stimulatory only in the concomitant presence of EL-BGDF-pik; (3) FLU-BA was slightly stimulatory when acting alone, but the addition of EL-BGDF-pik significantly raised the level of the response it elicited; (4) FLU-LPS and one FLU-Ficoll conjugate (FLU$_{150}$ Ficoll) were powerfully stimulatory over a 100,000-fold range of concentrations, and at no concentrations did BGDF affect their capacity. LPS or LPS + DXS at high concentrations behaved similarly. The FLU-Ficoll conjugate which exhibited BGDF independence was mitogenic at high concentrations, whereas the carrier AECM-Ficoll was not. Unconjugated BA was also mildly mitogenic in the presence of BGDF.

It should be noted that the degree of differentiation to anti-FLU AFC status was highest with the group 4 antigens (Table 4). Higher concentrations of FLU-LPS resulted in increased proliferation (up to 30%), but decreased specific antibody formation.[89] This latter observation would agree with Möller's[114] theory. Assay for ISC clones showed virtually all proliferating clones to be secreting antibody with all antigenic stimuli indicating that proliferation is accompanied by a degree of differentiation. The hapten-gelatin technique selects cells of a wide range of affinities for the hapten concerned.[5,6,87] The lower anti-FLU AFC clone frequency values could be due to low affinity cells being stimulated which would be detected with the indirect ISC plaque assay, as perhaps would smaller amounts of antibody, considering that a developing antiserum is used in the ISC test. Only 50% of the LPS + DXS-induced clones contained ISC. This level is significantly higher than reported by Kettman and Wetzel,[81,82] but comparison with antigen-driven results suggests that this system is more proliferation biased.

Table 4

BGDF DEPENDENCE OF RESPONSES ELICITED BY VARIOUS TI
STIMULI WHEN ACTING ON ISOLATED SINGLE FLU-ENRICHED B
CELLS IN VITRO[a]

		% FLU-enriched B cells responding to form			
		Proliferating clones[c]		Anti-FLU AFC clones[d]	
Stimulus	Putative class[b]	−BGDF[e]	+BGDF[f]	−BGDF	+BGDF
FLU-KLH	TD	<0.11	0.45	<0.11	<0.11
FLU-POL	TI-2	0.47 ± 0.26	9.3 ± 1.8	0.47 ± 0.47	5.5 ± 1.2
FLU$_{83}$Ficoll	TI-2	1.05 ± 0.29	15.9 ± 1.7	0.30 ± 0.13	6.0 ± 1.2
FLU-BA	TI-1	5.0 ± 2.7	15.7 ± 2.1	3.7 ± 0.07	9.4 ± 1.6
FLU-LPS	TI-1	15.2 ± 1.7	17.9 ± 1.9 (NS)	9.9 ± 1.6	9.3 ± 1.6
FLU$_{150}$Ficoll	TI-1	13.7 ± 1.7	15.9 ± 1.5 (NS)	9.0 ± 0.9	9.6 ± 1.7
LPS + DXS	Mitogen	25.1 ± 8.9	18.8 ± 6.9 (NS)	3.7 ± 0.7	4.2 ± 0.9

Note: Comprehensive results for each antigen have been published elsewhere.[89] The results as presented here represent an extracted summary of that data.

[a]   Responses generated from selected "antigenic" concentrations[89] of various TI stimuli when acting on single FLU-gelatin-enriched splenic B cells from normal adult CBA donors.
[b]   Classified as according to the literature.[21,102-104]
[c]   % B cells forming proliferating clones.
[d]   % B cells forming anti-FLU AFC clones assessed by plaque methodology.
[e]   Stimulation in the absence of EL-BGDF-pik.
[f]   Stimulation in the presence of EL-BGDF-pik.

These findings suggest a new classification of antigens to compete with the existing TI-1, TI-2, and TD classification. Antigens of group 1 could be considered as "classical TD", requiring the obligate interaction of MHC-restricted carrier-specific helper T cells. Antigens and mitogens of group 2 are "T cell-derived lymphokine-dependent" (LD), whereas those of group 3 are partially TI or LD and those of group 4 may eventually be proven to be totally TI (or "lymphokine-independent" [LI]).

Our data conflict with the interpretation of results generated in nonclonal assay systems recently reported in the literature. Responses to both TI-1 and TI-2 antigens have been shown to be dependent on T cell-derived factors in one study.[106] In another, it has been claimed that responses to both TNP-BA and TNP-Ficoll are dependent on T cells or an IL-2-like molecule derived from T cells,[104] and both antigens were nonstimulatory at low cell density (<2 × 10^5 cells per milliliter). Both these antigens act on single B cells in 10-$\mu\ell$ cultures in our hands (100 cells per milliliter), and the T cell-dependent component can be supplied by T cell-derived EL-BGDF-pik.[89] We have recently shown that human recombinant IL-2 can also promote the clonal growth of single B cells with both FLU-BA and a BGDF-dependent FLU-Ficoll conjugate.[93] Responses to TI-2 antigens have been claimed to be dependent on accessory cells.[108-110] Again, we have found no such dependence.

It is tempting to regard stimuli in group 4 as truly T-independent. However, the intentional addition of thymus filler cells to single B cells triggered by these agents

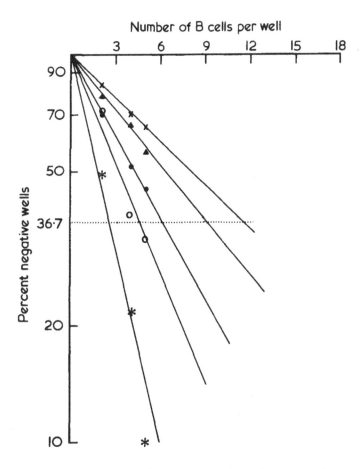

FIGURE 5. Limiting dilution analysis of the proliferation induced among FLU-gelatin fractionated splenic B cells by various TI stimuli. FLU-POL + BGDF (x—x); 0.01 μg/ml FLU-Ficoll (▲—▲); 0.1 μg/ml FLU-LPS (•—•); LPS + DXS (O—O); and 1 μg/ml FLU-Ficoll (*—*). BGDF was absent in all groups except where FLU-POL was the stimulus. The proliferative frequency values were as follows: FLU-POL + EL4-BGDF, 8.7% (6.2 to 11.1%); 0.01 μg/ml FLU_{150}Ficoll, 10.9% (8.1 to 13.7%); 0.1 μg/ml FLU-LPS, 16.2% (12.6 to 19.8%); LPS + DXS, 20.8% (16.5 to 25.2%); and 1 μg/ml FLU_{150}Ficoll, 38.9% (31.5 to 46.2%). The anti-FLU AFC clone frequencies (not shown) were as follows: for 0.1 μg/ml FLU-POL + BGDF, 4.2% (95% confidence limits 2.6 to 5.9%); 0.01 μg/ml FLU_{150}Ficoll, 6.4% (4.4 to 8.4%); 0.1 μg/ml FLU-LPS, 7.4% (5.2 to 9.6%); LPS + DXS, 3.9% (2.3 to 5.4%); and 1 μg/ml FLU_{150}-Ficoll, 5.3% (3.5 to 7.1%). (Reprinted from Pike, B. L. and Nossal, G. J. V., *J. Immunol.*, 136, 1687, 1984. With permission.)

markedly raised antibody formation.[6,89] This enhancement had been noted with FLU-POL + BGDF and LPS + DXS stimulation.[89] The anti-FLU AFC clone frequency values were raised approximately three- to fourfold with all stimuli and BGDF-thymus synergy was only noted with the group 2 antigen, namely, the BGDF-dependent FLU-POL and FLU_{53} Ficoll. Until the nature of this further "helper" effect is understood, it would be unwise to ignore the possibility that T cells or accessory cells are involved in some specific way. Equally possible is some nonspecific supportive effect leading to healthier cultures.

2. Comparative In Vitro Clone Formation of Various B Cell Subsets

The question regarding the comparative "clonability" of single B cells derived from normal adult CBA spleen, neonatal CBA spleen, and immunodeficient CBA/N (xid) spleen was addressed using FLU-gelatin binding B cells from each source and the various TI stimuli described above. FACS analyses have shown the FLU-enriched B cell populations from all three sources to be identical with respect to size distribution, total mIg avidity profile, and FLU-binding avidity profile.[91]

a. Behavior of CBA/N-Derived B Cells

CBA/N-derived B cells performed poorly with all stimuli used,[91] the maximum response being elicited from the CBA/N-derived FLU-enriched B cells by FLU-BA + BGDF (Table 5). Antigens classified in group 4 above (FLU-LPS, FLU$_{150}$Ficoll) and LPS + DXS were less effective, and those in group 2 above (FLU-POL) were essentially nonstimulatory. Thus, two typical supposed "TI-1" antigens, FLU-LPS and FLU-BA, which strongly stimulated normal adult single FLU-specific B cells to proliferate and form antibody, virtually failed to trigger CBA/N B cells of comparable antigen-binding avidity. Similar findings were noted with the mitogens LPS + DXS. The responses of unfractionated spleen cells from both CBA/H and CBA/N donors were also examined at a clonal level using the thymus "filler" cell-supported cloning system,[91] and again the CBA/N splenic B cells responded poorly, FLU-LPS and FLU-BA stimulating 1 in 2000 spleen cells from CBA donors to form anti-FLU AFC clones, compared to only 1 in 30,000 to 50,000 of CBA/N spleen cells.[91] With FLU-POL the CBA/N spleen cells were even less responsive.

It, therefore, emerges from these studies that CBA/N-derived FLU-enriched B cells are poorly responsive when stimulated with "TI-1" antigens (or mitogens) under conditions which elicit vigorous responses from normal CBA/H-derived FLU-enriched B cells which exhibit similar mIg and antigen-binding avidity profiles. It appears that CBA/N B cells have some defect which transcends the particular immunogenic properties of the stimuli used. Again, our results, obtained by studying the response at the clonal level, disagree with those of others[102,103] and cast some doubt on the categorization of antigens into "TI-1" and "TI-2" basis on their ability to elicit responses from CBA/N mice. However, while in our studies[91] only poor responses were elicited with TI-1 antigens, responsiveness to TI-2 stimuli (FLU-POL) was virtually nondetectable. There may thus be some genuine difference between the two types of antigen, with CBA/N mice exhibiting a partial defect toward one and a total defect to the second. The behavior of FLU-BA in this context again raises the point as to whether the dependence of the antigen on lymphokines for optimal stimulation is a better "categorization" parameter than the ability of the conjugate to elicit responses from various B cell subsets.

b. Behavior of Neonatal B Cells from Normal Mice

Similar studies were performed with single hapten-specific neonatal B cells (derived from normal 5- to 6-day-old CBA donors). These also yielded some surprising conclusions.[91] Both "TI-1" and "TI-2" stimuli caused adequate proliferation of from 2 to 5% of the hapten-specific B cells. FLU-LPS induced a higher proportion of B cells to form proliferating clones than either FLU-POL + BGDF or FLU-BA + BGDF. However, none of the antigens caused good antibody formation, probably because multivalent antigens can deliver signals impeding the differentiation of immature B cells.[31,112,113] As reported in other studies,[87] LPS + DXS stimulated the neonatal B cells to proliferate about as well as adult B cells (around 30%), but as seen with the antigenic stimuli, very little antibody formation was noted.[87,91]

Again, the behavior of single B cells from neonates does not support the notion that TI-1 antigens stimulate mature and immature B cells, but TI-2 stimulates only mature cells. The approach of monitoring both the proliferation and differentiation induced

Table 5

RESPONSES OF SINGLE HAPTEN-SPECIFIC B CELLS FROM ADULT CBA
AND CBA/N DONORS TO VARIOUS TI ANTIGENS[a]

| | % Proliferating clones | | % Anti-FLU-AFC clones | |
Stimulus	CBA/H	CBA/N	CBA/H	CBA/N
FLU-POL + BGDF	8.0 ± 1.6[b]	0.17 ± 0.04	3.7 ± 0.4	0.15 ± 0.04
FLU-BA + BGDF	19.0 ± 1.9	2.67 ± 0.74	12.2 ± 1.0	1.27 ± 0.15
FLU-BA	2.0	0.95	0.60	0.30
FLU-LPS (10 μg/mℓ)	18.9 ± 2.7	1.30 ± 0.4	9.5 ± 1.1	0.87 ± 0.42
FLU-LPS (0.1 μg/mℓ)	11.5 ± 2.3	0.52 ± 0.15	6.8 ± 0.6	0.34 ± 0.07
FLU$_{150}$Ficoll (1 μg/mℓ)	36.7 ± 2.4	1.77 ± 0.76	8.2 ± 1.7	1.15 ± 0.40
FLU$_{150}$Ficoll (0.01 μg/mℓ)	14.2 ± 2.1	0.50 ± 0.28	8.8 ± 1.1	0.38 ± 0.20
LPS + DXS	29.7 ± 5.8	2.33 ± 1.11	4.1 ± 0.5	0.16 ± 0.05

[a] Responses generated by FLU-gelatin binding B cells isolated from the spleens of age/sex matched adult CBA/H and CBA/N donors when stimulated by various TI antigens at limit dilution. FLU-POL was used at 0.1 μg/mℓ and FLU-BA at a 1 in 1000 dilution of stock solution. EL-BGDF-pik was used at 5% (v/v) of a × 10 concentrate where indicated. LPS (100 μg/mℓ) plus DXS (10 μg/mℓ) were used together as a mitogenic stimulus.

[b] Results represent the mean ± SEM of three to four experiments. Where no SEM is given results represent the mean of two experiments.

Reprinted from Nossal, G. J. V. and Pike, B. L., *J. Immunol.*, 132, 1696, 1984. With permission.

at the level of the single antigen-specific B cell should enlarge our understanding of the events of activation and help to establish a definitive classification for the antigens.

VI. SINGLE CELL APPROACH TO THE QUESTION OF SEPARATE TI- AND TD-RESPONSIVE B CELLS

A substantial body of evidence exists to suggest that the splenic B lymphocytes of the mouse can be divided operationally and phenotypically into two subpopulations, numerically approximately equal, but functionally distinct.[12,20-22] These two populations appear to differ in their requirements for triggering and have different alloantigenic markers. The predominant stream of experimental work attesting to these subsets comes from an analysis using the CBA/N xid mice, and of alloantisera prepared in them. Mice with the xid defect lack B cells bearing the determinants Lyb 3,[115] Lyb 5,[116] Lyb 7,[117] and Ia.W39.[118] The general pattern has been to divide these subsets into two distinct groups, namely, Lyb 5+ B cells, present in normal but not in CBA/N xid mice, and Lyb 5- B cells, present both in normal and CBA/N xid mice. Many of the conclusions drawn from studies on the functional differences between the two categories have been based either on studies where anti-Lyb 5 antibodies plus complement are used to eliminate the Lyb 5+ cells from normal spleen populations, or alternatively, studies using CBA/N-derived splenic B cells, i.e., Lyb 5- cells. There is a hidden assumption sometimes made that the Lyb 5- cells of CBA/N mice are identical to the Lyb 5- cells of normal mice, but our recent results[91] as discussed above cast doubt on the validity of this postulate. It has been suggested[12,21,22,119,120] that Lyb 5+ cells respond to antigen plus lymphokines of an antigenically nonspecific sort, whereas Lyb 5- cells obligatorily require an interaction of an MHC-restricted, antigen-specific kind with a helper T cell. Thus, Lyb 5- cells are putatively TD, whereas Lyb 5+ cells are, up to a point, TI.

Few clonal analyses on the existence of separable TI- and TD-responsive B cell subsets have been reported in the literature. In earlier studies in collaboration with Teale and McKenzie,[121] we separated B cells from normal mice on the basis of Lyb phenotype and enumerated clonable TI and TD B cells using haptenated-POL in the presence of thymus filler cells as the TI cloning system, and haptenated-KLH in the Klinman splenic microfocus assay[69,70] as the TD cloning system. These studies showed that none of the markers Lyb 1.1, Lyb 2.1, Lyb 3, Lyb 4.1, Lyb 5.2, or LyM 1 clearly demarcated TD from TI B cells. Lyb 5+ fractions, which on conventional wisdom should respond to TI antigens, were in fact 1.5-fold enriched for AFC precursors using a TD stimulus and actually slightly depleted for TI-stimulatable precursors. Lyb 5- B cell fractions, putatively TD, responded well in both the TD and TI systems, almost to the same degree as unfractionated B cells. These results are in conflict with the data of others as discussed above,[12,21,22,119,120] performed using nonclonal bioassay systems. Activated T cells are capable of secreting lymphokines[41-46] which are in turn capable of inducing polyclonal activation of B cells.[10,13,122-124] This lymphokine release could possibly occur in the Klinman assay system, although the low cloning frequencies generally reported[16,17,70-76] argue against this. In the TI, thymus filler system as previously used, i.e., without added lymphokines, it is possible that some form of "factor" release was affecting the responses.

A. Clonal Differentiation of Single B Hapten-Specific Cells in a TD Cloning System
1. TD Cloning System

A major area of controversy from our point of view has been whether separable distinct subsets of B cells exist which respond either to a TI or TD stimulus.[7,18,29-33] To date, clonal studies performed in our laboratory do not lend support to this view.[121] We have begun to readdress this issue, our interest restimulated both by the evolution of the new cloning methods and by the recent elucidation of the influences exerted by B cell-active lymphokines both on TI- and TD-induced differentiation of B cells.

We have used a TD cloning modified from that of Waldman et al.[77,78] to assess the responsiveness of FLU-enriched B cells to TD stimuli and to begin to address the question as to whether TI and TD B cell subsets exist.[80] Only antibody formation and not cellular proliferation could be assessed due to the presence of the T cells which obscured visualization. For the provision of carrier-specific help, we used small numbers (200 to 2000 T cells per 10-μl culture) of short-term (2- to 4-week) lines of T helpers derived from syngeneic, in vivo carrier-primed and boosted mice, the cells having been selected through stimulation by antigen, essentially according to the method of Augustin et al.[125] Short-term in vitro propagated helper T cells were selected for use on the basis of their higher efficiency, only 1500 T cells per 10 μl providing nonlimiting help compared to 2 to 4 × 10⁴/10 μl of in vivo-derived nylon wool nonadherent splenic T cells.

Comparative studies on the responses elicited from unfractionated spleen cells when stimulated in vitro with FLU-KLH in the presence of nonlimiting numbers showed both T cell populations supported a linear response. The frequency of anti-FLU AFC clones was higher in the presence of in vitro propagated T cells (1 in 3600 spleen cells) than in vivo-derived T cells (1 in 7000 spleen cells). The response in the absence of antigen was approximately one third lower with both forms of T cell help. These FLU-KLH-induced responses are similar to those reported by us for a FLU-POL-elicited response in the various TI cloning systems.[55-57] As so few T cells were used, B cell contamination was not a problem, thus avoiding the need for B cell depletion procedures.

2. Responses of Single Hapten-Specific B Cells to a TD Antigen

The frequency of anti-FLU AFC clone precursors among FLU-gelatin binding B cells was determined using this limiting dilution analysis. These studies have been reported in detail elsewhere.[80] Figure 6 shows the results of one experiment in which 1 in

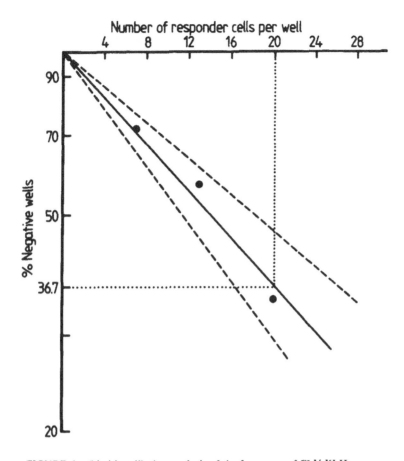

FIGURE 6. Limiting dilution analysis of the frequency of FLU-KLH responsive cells among FLU-gelatin binding B cells when cultured in the presence of 1500 short-term in vitro propagated helper T cells and 5×10^4 syngeneic irradiated spleen cells. The frequency of anti-FLU AFC precursors was calculated to be 4.92×10^{-2} (3.83 to 6.11%). In the presence of NIP-KLH, a frequency value of 2.14×10^{-2} (1.50 to 2.78%) was obtained (not shown). (Reprinted from Hebbard, G. S., Pike, B. L., and Nossal, G. J. V., *Proc. Natl. Acad. Sci. U.S.A.*, 81, 2479, 1984. With permission.)

20 FLU-specific B cells was shown to form an anti-FLU AFC clone in response to FLU-KLH. FLU-KLH stimulation of FLU-specific B cells in a series of 4 other experiments yielded a mean frequency of 3.65 ± 0.29% which is a value similar to that reported for a TI stimulus.[55-57] Thus, the mean enrichment factor obtained by the FLU-gelatin fractionation procedure is similar with TI and a TD stimulus. With NIP-KLH as the stimulus 1.81 ± 0.42% of the FLU-specific B cells formed anti-FLU PFC clones indicating a significant "bystander" effect. These values for FLU-KLH and NIP-KLH were significantly different (<0.0025).

3. Comparative Frequencies with TI and TD Stimuli

With the knowledge that the FLU-enriched B cell population responds with virtually equivalent efficiency to either a TI or a TD stimulus, the question of separable subsets was addressed using limiting dilution analysis. The frequency of anti-FLU AFC clone precursors was determined when FLU-specific B cells were stimulated in the "TD system" with FLU-POL acting as a TI stimulus and FLU-KLH as a TD stimulus.[80] The clone frequency differences between FLU-POL alone, or FLU-KLH alone, and FLU-POL and FLU-KLH acting together, provide little evidence for separate populations

Table 6
COMPARATIVE CLONING EFFICIENCIES OF HAPTEN-
SPECIFIC B CELLS WITH TD AND TI STIMULI[a]

Stimulus[b]	Anti-FLU AFC clone frequency ($\times 10^{-2}$)
FLU-KLH	3.33 ± 0.46[c]
NIP-KLH	1.70 ± 0.57
FLU-POL + NIP-KLH	8.63 ± 0.20
FLU-POL ± FLU-KLH	10.2 ± 0.52
FLU-POL	8.2[d]

[a] FLU-specific B cells were cultured at limit dilution in the presence of short-term in vitro propagated helper T cells and stimulated with the antigens as indicated.
[b] 50 ng/m*l* of FLU-KLH or NIP-KLH, 100 ng/m*l* of FLU-POL.
[c] Mean ± SEM of three separate experiments.
[d] Results of two experiments only.

Reprinted from Hebbard, G. S., Pike, B. L., and Nossal, G. J. V., *Proc. Natl. Acad. Sci. U.S.A.*, 81, 2479, 1984. With permission.

(Table 6). We have not assessed the Lyb status of the FLU-gelatin binding B cell population. Equivalent numbers of B cells bind from CBA/N donors as from normal donors.[91] By all other characteristics and surface markers which have been assessed, they are identical to those of normal mice,[91] and we presume the population to be representative of the total splenic B cell population per se.

4. Frequencies as Assessed by RIA

We have also assessed the cloning frequencies of FLU-specific B cells in this TD system using an antigen-specific sensitive radioimmunoassay (RIA) to detect antibody in the culture supernatant fluid. This work was performed in collaboration with Dr. J. J. Cebra. As reported above for TI stimuli in thymus-filler-supported cultures, the clone frequency values obtained were far above those assessed using the FLU-specific IgM plaque assay. Stimulation with FLU-KLH induced 40.3 ± 6.9% of single FLU-enriched B cells to form anti-FLU AFC clones. Bystander stimulation, as assessed using PC-KLH as a noncross-reacting hapten at the B cell level, induced 16.2 ± 7.0% clone formation. Stimulation of the same cell population with FLU-POL + EL-BGDF-pik in the presence of thymus filler cells yielded a clone frequency of 55% using the RIA assay procedure. If totally separate subsets exist, then each subset must react with a 100% cloning efficiency in each system. We know this cannot be the case, as presumably the Lyb phenotype is distributed among B cells of all sizes, and in the above TI system, the small cell subset of the FLU-enriched B cells contributes only about 20% of the total response.

VII. IN VITRO CLONAL GROWTH AND DIFFERENTIATION OF SINGLE B CELLS DERIVED FROM TOLERANT DONORS

The special sensitivity of immature, differentiating B lymphocytes to negative signaling by multivalent antigen has been best quantitated by in vitro clonal analysis.[8,68,87,113,126,127] Tolerance induced has been assessed by the reduction in the numbers of antigen-binding clonable antibody-forming cell (AFC) precursors (AFC-p) rather than as a reduction in total AFC or antibody produced, i.e., at the level of the target cell per se.

We have documented[87,126,127] the existence of a population of antigen-specific B cells within an animal that has been rendered tolerant *in utero* or neonatally with the following characteristics: normal mIg receptor coat, normal antigen avidity profile, normal size distribution, and normal number, using the hapten-gelatin technique to isolate hapten-specific B cells from the spleens of tolerant donors and subsequent FACS analysis. These cells were unable to generate AFC clones.[126,127]

The ability to examine both the proliferative and differentiative capacity of the single "anergic" B cells to a variety of TI triggering stimuli, in isolation from potential suppressive or enhancing influences of other cell types, provided an approach to further understanding of B cell tolerance. We were able to study the behavior of single anergic B cells on interaction with a variety of immunogenic stimuli and determine whether the lesion in the cell lay in its ability to undergo blastogenesis and clonal expansion, and/or its inability to differentiate to AFC status. The responsiveness from normal control donors was compared with those from donors tolerized by administration of FLU-human gammaglobulin in the perinatal period in each of the cloning systems.[87] The normalized results of these studies are shown in Table 7.

Both the proliferative and differentiative capacity of the anergic FLU-specific B cells were diminished when EL-BGDF-dependent antigenic stimulus was used. With BGDF-independent stimuli, the proliferative capacity was slightly affected, but the anti-FLU AFC clone-developing capacity was severely affected. This finding delineates anergic B cells from other "nonresponsive" B cell subsets such as those derived from CBA/N donors.

One could speculate as to whether the lesion in the anergic B cell is the lack of cell surface receptors for B cell growth and differentiation factor(s), or whether the negative signal delivered to the cell on contact with toleragenic antigen resulted in an intrinsic inability to respond to a physiological, antigenic stimulus. The uncoupling of the lesion into one of growth and/or differentiation is possible in the systems we describe with the imminent availability of purified sources of factors which exert control over each of these processes separately. The question could then be asked as to whether the anergic B cell could in fact undergo proliferation but not differentiation, and if so, whether the addition of differentiation-promoting factors (should such exist) would have any effect on continued proliferation or perhaps lead to the death of the cells.

VIII. CONCLUSIONS

The recent development of more refined approaches using enriched B cell populations and single B cell cloning systems[40,45,81,82,84,86] or low density cultures[44,46,128,129] to assess the events following activation of the resting G_0 B cell into the pathway of division and maturation has clearly shown that these events can be considered as distinct entities of the immune response.

Analysis of the cell cycle status has shown 95% of murine splenic B cells to be in G_0.[14,15] Multivalent antigen or anti-Ig antibodies alone cannot trigger the resting B cell past the G_1 state into division and differentiation, and additional signals are required for the progression of the response.[45-49]

For activation of resting B cells to antibody-forming status by multivalent TI antigens, or anti-Ig antibodies, there is now substantial evidence both from our own studies[40,45,89] and those of others[44,86] that additional B cell growth and differentiation factors are required to both initiate proliferation and subsequent antibody formation. These factors, produced by mitogenic stimulation of cells from cloned T cell lines or hybridomas, appear to act in a nonantigen-specific, non-MHC-restricted manner.

The conclusion emerging[42,43,130-132] is that B lymphocyte growth and differentiation may require the simultaneous or sequential cooperative action of two or more growth

Table 7

CLONING CAPACITY OF SINGLE FLU-ENRICHED B CELLS FROM TOLERANT DONORS WHEN STIMULATED IN VITRO WITH VARIOUS TI STIMULI[a]

Stimulus	Proliferation	% Control frequency value		
		p Value[b]	Differentiation	p Value[b]
FLU-POL + BGDF[c]	23 ± 3	<0.0005	5.4 ± 1	<0.0025
FLU-BA + BGDF	25 ± 1	<0.0025	2.5 ± 1	<0.0005
FLU-LPS				
10 μg/ml	73 ± 3	<0.01	13 ± 5	<0.05
0.1μg/ml	65 ± 13	NS	11 ± 2	<0.0025
FLU$_{150}$Ficoll				
1 μg/ml	97 ± 11	NS	24 ± 9	<0.0005
0.01 μg/ml	88 ± 30	NS	14 ± 3	<0.005
LPS + DXS[3]	98 ± 12	NS	22 ± 6	<0.05

[a]   FLU-gelatin binding B cells were cultured at limit dilution in the filler cell-free cloning systems with the antigenic stimuli as indicated. Values shown are normalized to the control frequency value within each experiment and represent mean ± SEM of four to eight experiments.

[b] p value, as determined by student's t test.

[c]   These data represent summary data from a series of experiments, some of which have been published in detail elsewhere.[89]

regulators, e.g., growth factors together with differentiation factors. The issue is made more complex by the possibility that different B cell subsets respond to different sets of signals. Evidence exists for an initial requirement for BCGF for the induction to proliferating status, some authors claiming two distinct BCGFs to be needed.[130,131] Differentiation-promoting factors or T cell-replacing factors (TRF) have been claimed to be required for progression to AFC status.[132,133] Many of these proposals have been based on analyses performed using complex bioassay systems, making the possibility of uncoupling the effects on proliferation from those on differentiation extremely difficult.

Our own work so far, using single hapten-specific B cells as the unequivocal target cell has failed to reveal evidence for the existence of a pure differentiation-inducing factor, e.g., one capable of driving a B cell into an antibody-forming cell without intervening division. When purified factor sources become available, growth-promoting factors as distinct from differentiation-promoting, this question can be addressed. Our studies to date with various purified lymphokines indicate that most if not all factors active on B cells promote both proliferation and differentiation.

Activation of resting B cells to antibody-forming status by soluble TD antigens has been shown by many to require two further signals in addition to antigen, as proposed by Schimpl and Wecker.[36] TD antigen alone cannot induce B cells into blastogenesis and proliferation. For induction to at least proliferation status, there is evidence for a requirement for an antigen-specific, MHC-restricted, T cell-derived helper signal either delivered by antigen-specific T cells or factors derived from antigen-specific T cells.[13,19,95,128,129,132,134,135] For maturation of the activated blast cells to occur, there is evidence for a requirement for nonantigen-specific, non-MHC-restricted, T cell-derived factors.[10,13,95,96,122,128,132,134] Submitogenic concentrations of the B cell mitogen, lipopolysaccharide (LPS), have been reported to be able to substitute for the second specific signal[134] and accessory cells have been reported to be able to substitute for the third nonspecific signal.[128] A key issue which we hope to address as we perfect the cloning of single hapten-specific B cells with TD antigens and T cells is whether T cell help is simply a particularly effective method of delivering a conventional growth and differentiation factor.

The stage is now set for a much more meaningful analysis of the cellular and molecular events involved in B lymphocyte triggering. Recombinant DNA technology will soon provide the family of relevant lymphokines in pure form. Knowledge about receptors for B cell-active factors will follow shortly thereafter. This, together with the availability of a variety of stimulatory systems for B cell clonal proliferation, will soon give us a better knowledge of B cell subsets. Armed with a good appreciation of B cell subsets, B cell growth and differentiation factors and receptors for them, and refined "TI" and "TD" triggering modalities, the nature of the lesion in the tolerant B cell will soon be much better defined. The next phase will then be to come to grips with the mechanisms of gene activation at the molecular level.

ACKNOWLEDGMENTS

We wish to acknowledge the collaboration of Drs. J. J. Cebra, J. W. Schrader, F. L. Battye, I. Clark-Lewis, D. L. Vaux, and G. S. Hebbard in some of the original work presented herein. The technical assistance of A. Milligan, L. Gibson, A. Mason, M. Zanoni, and M. Ludford is gratefully acknowledged.

This original work was supported by the National Health and Medical Research Council, Canberra, Australia; by Grant Number AI-03958 from the National Institute of Allergy and Infectious Diseases, U.S. Public Health Service; and by the generosity of a number of private donors to The Walter and Eliza Hall Institute.

REFERENCES

1. Osmond, D. G., Generation of B lymphocytes in the bone marrow, in *B Lymphocytes in the Immune Response*, Cooper, M., Mosier, D., Scher, I., and Vitetta, E. S., Eds., Elsevier/North-Holland, New York, 1979, 63.
2. Kincade, P. W., Formation of B lymphocytes in fetal and adult life, *Adv. Immunol.*, 31, 177, 1981.
3. Scher, I., B lymphocyte ontogeny, *CRC Crit. Rev. Immunol.*, 1, 287, 1981.
4. Haas, W. and Layton, J. E., Separation of antigen-specific lymphocytes. I. Enrichment of antigen-binding cells, *J. Exp. Med.*, 141, 1004, 1975.
5. Nossal, G. J. V., Pike, B. L., and Battye, F. L., Sequential use of hapten-gelatin fractionation and fluorescence-activated cell sorting in the enrichment of hapten-specific B lymphocytes, *Eur. J. Immunol.*, 8, 151, 1978.
6. Pike, B. L., Vaux, D. L., and Nossal, G. J. V., Single cell studies on hapten-specific B lymphocytes: differential cloning efficiency of cells of various sizes, *J. Immunol.*, 131, 544, 1983.
7. Layton, J. E., Pike, B. L., Battye, F. L., and Nossal, G. J. V., Cloning of B cells positive or negative for surface IgD. I. Triggering and tolerance in T-independent system, *J. Immunol.*, 123, 702, 1979.
8. Mosier, D. E., Mond, J. J., and Goldings, E. A., The ontogeny of thymic independent antibody responses in vitro in normal mice and mice with an X-linked B cell defect, *J. Immunol.*, 119, 1874, 1977.
9. Mosier, D. E., Zitron, I., Mond, J. J., Ahmed, A., Scher, I., and Paul, W. E., Surface IgD as a functional receptor for a subclass of B lymphocytes, *Immunol. Rev.*, 37, 89, 1977.
10. Ratcliffe, M. J. H. and Julius, M. H., H-2-restricted T-B interactions involved in polyspecific B cell responses mediated by soluble antigen, *Eur. J. Immunol.*, 12, 634, 1982.
11. Andersson, J., Lernhardt, W., and Melchers, F., The purified protein derivative of tuberculin. A B-cell mitogen that distinguishes in its action resting, small B cells from activated B-cell basis, *J. Exp. Med.*, 150, 1339, 1979.
12. Asano, Y., Singer, A., and Hodes, R. J., Role of the major histocompatibility complex in T cell activation of B cell subpopulations. MHC restricted and unrestricted B cell responses are mediated by distinct B cell subpopulations, *J. Exp. Med.*, 154, 1100, 1981.

13. Julius, M. H., von Boehmer, H., and Sidman, C. L., Dissociation of two signals required for activation of resting B cells, *Proc. Natl. Acad. Sci. U.S.A.*, 79, 1989, 1982.

14. Monroe, J. G. and Cambier, J. C., Cell cycle dependence for expression of membrane associated IgD, IgM and Ia antigen on mitogen stimulated B lymphocytes, *Ann. N.Y. Acad. Sci.*, 399, 238, 1982.

15. Monroe, J. G., Havran, W. L., and Cambier, J. C., Enrichment of viable lymphocytes in defined cycle phases by sorting on the basis of pulse width of axial light extinction, *Cytometry*, 3, 24, 1982.

16. Press, J. L., Strober, S., and Klinman, N. R., Characterization of B cell subpopulations by velocity sedimentation, surface Ia antigens and immune function, *Eur. J. Immunol.*, 7, 329, 1977.

17. Teale, J. M., Howard, M. C., Falzon, E., and Nossal, G. J. V., B lymphocyte subpopulations separated by velocity sedimentation, *J. Immunol.*, 121, 2554, 1978.

18. Gorczynski, R. and Feldman, M., B cell heterogeneity. Differences in size of B lymphocytes responding to T-dependent and T-independent antigens, *Cell Immunol.*, 18, 88, 1975.

19. Melchers, F., Andersson, J., Lernhardt, W., and Schreier, M. H., H-2-unrestricted polyclonal maturation without replication of small B cells induced by antigen-activated T cell help factors, *Eur. J. Immunol.*, 10, 679, 1980.

20. Hodes, R. J., Shigeta, M., Hathcock, K. S., Fathman, C. G., and Singer, A., Role of the major histocompatibility complex in T cell activation of B cell subpopulations: antigen-specific and H-2-restricted monoclonal T_H cells activate Lyb-5$^+$ B cells through an antigen-nonspecific and H-2-unrestricted effector pathway, *J. Immunol.*, 129, 267, 1982.

21. Mond, J. J., Use of T lymphocyte regulated type 2 antigens for the analysis of responsiveness of Lyb 5$^+$ and Lyb 5$^-$ B lymphocytes to T lymphocyte derived factors, *Immunol. Rev.*, 64, 99, 1982.

22. Tominaga, A., Takatsu, K., and Hamaoka, T., Antigen-induced T cell-replacing factor. II. X-linked gene control for the expression of TRF acceptor site(s) on B lymphocytes and preparation of specific antiserum to that acceptor, *J. Immunol.*, 124, 2423, 1980.

23. Vitetta, E. S., Melcher, U., McWilliams, M., Phillips-Quagliata, J., Lamm, M., and Uhr, J. W., Cell surface immunoglobulin. XI. The appearance of an IgD-like molecule on murine lymphoid cells during ontogeny, *J. Exp. Med.*, 141, 206, 1975.

24. Scott, D. W., Venkataraman, M., and Jandinski, J. J., Multiple pathways of B lymphocyte tolerance, *Immunol. Rev.*, 43, 241, 1979.

25. Coffman, R. L. and Cohn, M., The class of surface Ig on virgin and memory B lymphocytes, *J. Immunol.*, 118, 1806, 1977.

26. Zan-Bar, I., Vitetta, E. S., Assisi, F., and Strober, S., The relationship between surface immunoglobulin isotype and immune function of murine B lymphocytes, *J. Exp. Med.*, 147, 1374, 1978.

27. Scott, D. W., Layton, J. E., and Nossal, G. J. V., Role of IgD in the immune response and tolerance, *J. Exp. Med.*, 146, 1473, 1977.

28. Layton, J. E., Johnson, G. R., Scott, D. W., and Nossal, G. J. V., The antidelta suppressed mouse, *Eur. J. Immunol.*, 8, 325, 1978.

29. Quintans, J. and Cosenza, H., Antibody response to phosphorylcholine *in vitro*. II. Analysis of T-dependent and T-independent responses, *Eur. J. Immunol.*, 6, 399, 1976.

30. Jennings, J. J. and Rittenberg, M. B., Evidence for separate subpopulations of B cells responding to T-dependent and T-independent immunogens, *J. Immunol.*, 117, 1749, 1976.

31. Layton, J. E., Teale, J. M., and Nossal, G. J. V., Cloning of B cells positive or negative for surface IgD. II. Triggering and tolerance in the T-dependent splenic focus assay, *J. Immunol.*, 123, 709, 1979.

32. Lewis, G. K., Ranken, R., Nitecki, D. E., and Goodman, J. W., Murine B cell subpopulations responsive to T-dependent and T-independent antigens, *J. Exp. Med.*, 144, 382, 1976.

33. Fung, J. and Kohler, H., Immune response to phosphorylcholine. VII. Functional evidence for three separate B cell subpopulations responding to TI and TD PC-antigens, *J. Immunol.*, 125, 640, 1980.

34. Dutton, R. W., Falkoff, R., Hirst, J. A., Hoffman, M., Kappler, J. W., Kettman, J. R., Lesley, J. F., and Vann, D., Is there evidence for a nonantigen specific diffusable chemical mediator from the thymus-derived cell in the initiation of the immune response? *Prog. Immunol.*, 1, 355, 1971.

35. Hoffman, M. K., Antibody regulates the cooperation of B cells with helper cells, *Immunol. Rev.*, 49, 79, 1980.

36. Schimpl, A. and Wecker, E., A third signal in B cell activation given by TRF, *Transplant. Rev.*, 23, 176, 1975.

37. Watson, J., Gillis, S., Marbrook, J., Mochizuki, D., and Smith, K. A., Biochemical and biological characterization of lymphocyte regulatory molecules. I. Purification of a class of murine lymphokines, *J. Exp. Med.*, 150, 849, 1979.

38. Parker, D. C., Induction and suppression of polyclonal antibody responses by anti-Ig reagents and antigen-non specific helper factors, *Immunol. Rev.*, 52, 115, 1980.

39. Takatsu, K., Tanaka, K., Tominaga, A., Kumahara, Y., and Hamaoka, T., Antigen-induced T cell-replacing factor (TRF). III. Establishment of T cell hybrid clone continuously producing TRF and functional analysis of released TRF, *J. Immunol.*, 125, 2646, 1980.
40. Vaux, D. L., Pike, B. L., and Nossal, G. J. V., Antibody production by single, hapten-specific B lymphocytes: an antigen-driven cloning system free of filler or accessory cells, *Proc. Natl. Acad. Sci. U.S.A.*, 78, 7702, 1981.
41. Pure, E., Isakson, P. C., Takatsu, K., Hamada, T., Swain, S. L., Dutton, R. W., Dennert, G., Uhr, J. H., and Vitetta, E. S., Induction of B cell differentiation by T cell factors. Stimulation of IgM secretion by products of a T cell hybridoma and a T cell line, *J. Immunol.*, 127, 1953, 1981.
42. Marrack, P., Graham, S. D., Kushnir, E., Leibson, H. J., Boehm, N., and Kappler, J. W., Nonspecific factors in B cell responses, *Immunol. Rev.*, 63, 33, 1982.
43. Swain, S. L., Wetzel, G. D., Saubiran, P., and Dutton, R. W., T cell replacing factors in the B cell response to antigen, *Immunol. Rev.*, 63, 111, 1982.
44. Farrar, J. J., Benjamin, W. R., Hilfiker, M., Howard, M., Farrar, W. L., and Fuller-Farrar, J., The biochemistry, biology and role of interleukin 2 in the induction of cytotoxic T cell and antibody-forming B cell responses, *Immunol. Rev.*, 63, 129, 1982.
45. Pike, B. L., Vaux, D. L., Clark-Lewis, I., Schrader, J. W., and Nossal, G. J. V., Proliferation and differentiation of single, hapten-specific B lymphocytes promoted by T cell factor(s) distinct from T cell growth factor, *Proc. Natl. Acad. Sci. U.S.A.*, 79, 6350, 1982.
46. Howard, M., Farrar, J., Hilfiker, M., Johnson, B., Takatsu, K., Hamaoka, T., and Paul, W. E., Identification of a T cell-derived B cell growth factor distinct from interleukin 2, *J. Exp. Med.*, 155, 914, 1982.
47. Cambier, J. C., Monroe, J. G., and Neale, M. J., Definition of conditions that enable antigen-specific activation of the majority of isolated trinitrophenol-binding B cells, *J. Exp. Med.*, 156, 1635, 1982.
48. Snow, E. C., Vitetta, E. S., and Uhr, J. W., Activation of antigen-enriched B cells. I. Purification and responses to thymus-independent antigens, *J. Immunol.*, 130, 614, 1983.
49. DeFranco, A., Raveche, E., Asofsky, R., and Paul, W. E., Frequency of B lymphocytes responsive to anti-immunoglobulin, *J. Exp. Med.*, 155, 1523, 1982.
50. Jones, B. and Janeway, C. A., Cooperative interaction of B lymphocytes with antigen-specific helper T lymphocytes is MHC restricted, *Nature (London)*, 292, 547, 1981.
51. Katz, D. H., Hamaoka, T., and Benacerraf, B., Cell interactions between histoincompatible T and B lymphocytes. II. Failure of physiologic cooperative interactions between T and B lymphocytes from allogenic donor strains in humoral response to hapten-protein conjugates, *J. Exp. Med.*, 137, 1405, 1973.
52. Nisbet-Brown, E., Singh, B., and Diener, E., Antigen recognition. V. Requirement for histocompatibility between antigen-presenting cell and B cell in the response to a thymus-dependent antigen, and lack of allogeneic restriction between T and B cells, *J. Exp. Med.*, 154, 676, 1981.
53. Mishell, R. I. and Dutton, R. W., Immunization of dissociated spleen cell cultures from normal mice, *J. Exp. Med.*, 126, 423, 1967.
54. Marbrook, J. and Haskill, J. S., The in vitro response to sheep erythrocytes by mouse spleen cells. Segregation of distinct events leading to antibody formation, *Cell. Immunol.*, 13, 12, 1974.
55. Stocker, J. W., Estimation of hapten-specific antibody forming cell precursors in microculture, *Immunology*, 30, 181, 1976.
56. Nossal, G. J. V. and Pike, B. L., Single cell studies on the antibody-forming potential of fractionated, hapten-specific B lymphocytes, *Immunology*, 30, 189, 1976.
57. Nossal, G. J. V. and Pike, B. L., Improved procedures for the fractionation and in vitro stimulation of hapten-specific B lymphocytes, *J. Immunol.*, 120, 145, 1978.
58. Andersson, J., Coutinho, A., Lernhardt, W., and Melchers, F., Clonal growth and maturation to immunoglobulin secretion in vitro of every growth-inducible B lymphocyte, *Cell*, 10, 27, 1977.
59. Andersson, J., Coutinho, A., and Melchers, F., Frequencies of mitogen-reactive B cells in the mouse. I. Distribution in different lymphoid organs from different inbred strains of mice at different ages, *J. Exp. Med.*, 145, 1511, 1977.
60. Andersson, J., Coutinho, A., and Melchers, F., Frequencies of mitogen-reactive B cells in the mouse. II. Frequencies of B cells producing antibodies which lyse sheep or horse erythrocytes, and trinitrophenylated or nitroiodophenylated sheep erythrocytes, *J. Exp. Med.*, 145, 1520, 1977.
61. Pillai, P. S. and Scott, D. W., Cellular events in tolerance. IX. Maintenance of immunological tolerance in the presence of normal B-cell precursors and in the absence of demonstrable suppression, *Cell. Immunol.*, 77, 69, 1983.
62. Pillai, P. S., Scott, D. W., Piper, M., and Corley, R. B., Effect of recent antigen exposure on the functional expression of B cell populations, *J. Immunol.*, 129, 1023, 1982.
63. Metcalf, D., Role of mercaptoethanol and endotoxin in stimulating B lymphocyte colony formation *in vitro*, *J. Immunol.*, 116, 635, 1976.

64. Metcalf, D., Nossal, G. J. V., Warner, N. L., Miller, J. F. A. P., Mandel, T. E., Layton, J. E., and Gutman, G., Growth of B lymphocyte colonies *in vitro, J. Exp. Med.*, 142, 1534, 1975.

65. Kincade, P. W., Paige, C. J., Parkhouse, R. M. E., and Lee, G., Characterization of murine colony-forming B cells. I. Distribution, resistance to anti-immunoglobulin antibodies, and expression of Ia antigens, *J. Immunol.*, 120, 1289, 1978.

66. Kurland, J. I., Kincade, P. W., and Moore, M. A. S., Regulation of B lymphocyte clonal proliferation by stimulatory and inhibitory macrophage-derived factors, *J. Exp. Med.*, 146, 1420, 1977.

67. Scott, D. W., Layton, J. E., and Johnson, G. R., Surface immunoglobulin phenotype of murine B cells which form B cell colonies in agar, *Eur. J. Immunol.*, 8, 286, 1978.

68. Nossal, G. J. V., Pike, B. L., and Battye, F. L., Mechanisms of clonal abortion tolerogenesis. II. Clonal behaviour of immature B cells following exposure to anti-μ chain antibody, *Immunology*, 37, 203, 1979.

69. Klinman, N. R., The mechanisms of antigenic stimulation of primary and secondary clonal precursor cells, *J. Exp. Med.*, 136, 241, 1972.

70. Klinman, N. R. and Press, J..L., The B cell specificity repertoire: its relationship to definable subpopulations, *Transplant. Rev.*, 24, 41, 1975.

71. Press, J. L. and Klinman, N. R., Frequency of hapten-specific B cells in neonatal and adult mouse spleens, *Eur. J. Immunol.*, 4, 155, 1974.

72. Hurwits, J. L., Tagart, V. B., Schweitzer, P. A., and Cebra, J. J., Patterns of isotype expression by B cell clones responding to thymus-dependent and thymus independent antigens in vitro, *Eur. J. Immunol.*, 12, 342, 1982.

73. Cebra, J. J., Cebra, E. R., Clough, E. R., Fuhrman, J. A., and Schweitzer, P. A., Relationships and regulation of IgG, IgE or IgA expression during clonal outgrowth of primed B cells, in *Regulation of Immune Response, 8th International Convocation on Immunology*, Peary, P. L. and Jacobs, D. M., Eds., A. M. Herst, New York, 1982, 107.

74. Cebra, J. J., Komisar, J. L., and Schweitzer, P. A., C_H isotype "switching" during normal B cell development, *Ann. Rev. Immunol.*, 2, 493, 1984.

75. Teale, J. M., Howard, M. C., and Nossal, G. J. V., B lymphocyte subpopulations separated by velocity sedimentation. II. Characterization of tolerance susceptibility, *J. Immunol.*, 121, 2561, 1978.

76. Teale, J. M., Lafrenz, D., Klinman, N. R., and Strober, S., Immunoglobulin class commitment exhibited by B lymphocytes separated according to surface isotype, *J. Immunol.*, 126, 1952, 1981.

77. Waldman, H., Lefkovits, I., and Quintáns, J., Limiting dilution analysis of helper T-cell function, *Immunology*, 28, 1135, 1975.

78. Quintans, J. and Lefkovits, I., Precursor cells specific to sheep red cells in nude mice. Estimation of frequency in the microculture system, *Eur. J. Immunol.*, 3, 392, 1973.

79. Teale, J. M., The use of specific helper T cell clones to study the regulation of isotype expression by antigen-stimulated B cell clones, *J. Immunol.*, 131, 2170, 1983.

80. Hebbard, G. S., Pike, B. L., and Nossal, G. J. V., Single cell studies on hapten-specific B cells: response to T-cell-dependent antigens, *Proc. Natl. Acad. Sci. U.S.A.*, 81, 2479, 1984.

81. Kettman, J. and Wetzel, M., Antibody synthesis in vitro, a marker of B cell differentiation, *J. Immunol. Methods*, 39, 203, 1980.

82. Wetzel, G. D. and Kettman, J. R., Activation of murine B lymphocytes. III. Stimulation of B lymphocyte clonal growth with lipopolysaccharide and dextran sulfate, *J. Immunol.*, 126, 723, 1981.

83. Gronowicz, E., Coutinho, A., and Melchers, F., A plaque assay for cell secreting immunoglobulin of a given type or class, *Eur. J. Immunol.*, 6, 588, 1976.

84. Wetzel, G. D., Swain, S. L., and Dutton, R. W., A monoclonal T cell replacing factor, (DL)TRF, can act directly on B cells to enhance clonal expansion, *J. Exp. Med.*, 156, 306, 1982.

85. Swain, S. L. and Dutton, R. W., Production of a B cell growth-promoting activity, (DL)BCGF, from a cloned T cell line and its assay on the BCL_1 B cell tumor, *J. Exp. Med.*, 156, 1821, 1982.

86. Swain, S., Dennert, G., Warner, J., and Dutton, J., Culture supernatant of a stimulated T cell line have helper activity that synergizes with IL-2 in the response of B cells to antigen, *Proc. Natl. Acad. Sci. U.S.A.*, 78, 2517, 1981.

87. Pike, B. L., Abrams, J., and Nossal, G. J. V., Clonal anergy: inhibition of antigen-driven proliferation among single B lymphocytes from tolerant animals, and partial breakage of anergy by mitogen, *Eur. J. Immunol.*, 13, 214, 1983.

88. Pike, B. L., Jennings, G., and Shortman, K., A simple semi-automated plaque method for the detection of antibody forming cell clones in microcultures, *J. Immunol. Methods*, 52, 25, 1982.

89. Pike, B. L. and Nossal, G. J. V., A reappraisal of "T-independent" antigens. I. Effect of lymphokines on the response of single adult hapten-specific B lymphocytes, *J. Immunol.*, 132, 1687, 1984.

90. Paul, W. E., Nomenclature for B cell stimulatory factors (Kyoto, 1983), *Eur. J. Immunol.*, 13, 956, 1983.

91. Nossal, G. J. V. and Pike, B. L., A reappraisal of "T-independent" antigens. II. Studies on single, hapten-specific B cells from neonatal CBA/H or CBA/N mice fail to support classification into TI-1 and TI-2 categories, *J. Immunol.*, 132, 1696, 1984.

92. Mond, J. J., Mongini, P. K. A., Sieckmann, D., and Paul, W. E., Role of T lymphocytes in the response to TNP-AECM-Ficoll, *J. Immunol.*, 125, 1066, 1980.

93. Pike, B. L., Raubitschek, A., and Nossal, G. J. V., Human interleukin 2 can promote the growth and differentiation of single hapten-specific B cells in the presence of specific antigen, *Proc. Natl. Acad. Sci. U.S.A.*, 81, 7917, 1984.

94. Nossal, G. J. V. and Pike, B. L., Single cell studies on the activation requirements of B lymphocyte subsets, in *Progress in Immunology: Proceedings of the 5th International Congress in Immunology*, Tada, T., Ed., Academic Press, Japan, 1983, 701.

95. Andersson, J. and Melchers, F., T cell-dependent activation of resting B cells: requirement for both nonspecific unrestricted and antigen-specific Ia-restricted soluble factors, *Proc. Natl. Acad. Sci. U.S.A.*, 78, 2497, 1981.

96. Andersson, J., Schreier, M. H., and Melchers, F., T-cell-dependent B-cell stimulation is H-2 restricted and antigen dependent only at the resting B-cell level, *Immunology*, 77, 1612, 1980.

97. Miller, J. F. A. P., Lymphocyte interactions in antibody responses, *Int. Rev. Cytol.*, 33, 77, 1972.

98. Davis, A. J. S., Carter, R. L., Leuchars, E., Wallis, V., and Dietrich, F. M., The morphology of immune reactions in normal, thymectomized and reconstituted mice. III. Response to bacterial antigens: salmonella flagella antigen and pneumococcal polysaccharide, *Immunology*, 19, 945, 1970.

99. Humphrey, J. H., Parrott, D. M. V., and East, J., Studies on globulin and antibody production in mice thymectomized at birth, *Immunology*, 7, 419, 1964.

100. Feldmann, M., Induction of immunity and tolerance in vitro. I. The relationship between degree of hapten conjugation and the immunogenicity of dinitrophenylated polymerized flagellin, *J. Exp. Med.*, 135, 735, 1972.

101. Scher, I., The CBA/N mouse strain. An experimental model illustrating the influence of the X-chromosome on immunity, *Adv. Immunol.*, 33, 1, 1982.

102. Mond, J. J., Scher, I., Mosier, D. E., Baese, M., and Paul, W. E., T-independent responses in B cell-defective CBA/N mice to Brucella abortus and to trinitrophenyl (TNP) conjugates of Brucella abortus, *Eur. J. Immunol.*, 8, 459, 1978.

103. Mosier, D. E., Scher, I., and Paul, W. E., In vitro responses of CBA/N mice: spleen cells of mice with an X-linked defect that precludes immune responses to several thymus independent antigens can respond to TNP-lipopolysaccharide, *J. Immunol.*, 117, 1363, 1976.

104. Mond, J. J., Farrar, J., Paul, W. E., Fuller-Farrar, J., Schaeffer, M., and Howard, M., T cell dependence and factor reconstitution of in vitro antibody responses to TNP-B. abortus and TNP-Ficoll: restoration of depleted responses with chromatographed fractions of a T cell-derived factor, *J. Immunol.*, 131, 633, 1983.

105. Nordin, A. A. and Schreier, M. H., T cell control of the antibody response to the T-independent antigen, DAGG-Ficoll, *J. Immunol.*, 129, 557, 1982.

106. Enders, R. O., Kushnir, E., Kappler, J. W., Marrack, P., and Knisky, S. C., A requirement for nonspecific T cell factors in antibody responses to "T cell independent" antigens, *J. Immunol.*, 130, 781, 1983.

107. Yaffe, L. J., Mond, J. J., Ahmed, A., and Scher, I., Analysis of the B cell subpopulations by allogeneic effect factor. I. MHC restricted enhancement of B cell responses to thymic-independent antigens. Types 1 and 2, in normal and CBA/N mice, *J. Immunol.*, 130, 632, 1983.

108. Chused, T. M., Kassan, S. S., and Mosier, D. E., Macrophage requirement for the in vitro response to TNP-Ficoll: a thymic-independent antigen, *J. Immunol.*, 116, 1579, 1976.

109. Morrisey, P. J., Boswell, H. S., Scher, I., and Singer, A., Role of accessory cells in B cell activation. IV. Accessory cells are required for the in vitro generation of thymic independent type 2 antibody responses to polysaccharide antigens, *J. Immunol.*, 127, 1345, 1981.

110. Letvin, N. L., Benacerraf, B., and Germain, R. N., B lymphocyte responses to trinitrophenyl-conjugated Ficoll: requirement for T lymphocytes and Ia-bearing adherent cells, *Proc. Natl. Acad. Sci. U.S.A.*, 78, 5113, 1981.

111. Mond, J. J., Mongini, P. K. A., Sieckmann, D., and Paul, W. E., Role of T lymphocytes in the response to TNP-AECM-Ficoll, *J. Immunol.*, 125, 1066, 1980.

112. Nossal, G. J. V. and Pike, B. L., Mechanisms of clonal abortion tolerogenesis. I. Response of immature hapten-specific B lymphocytes, *J. Exp. Med.*, 148, 1161, 1978.

113. Metcalf, E. S., Schrater, A. F., and Klinman, N. R., Murine models of tolerance induction in developing and mature B cells, *Immunol. Rev.*, 43, 143, 1979.

114. Möller, G., One non-specific signal triggers B lymphocytes, *Transplant. Rev.*, 23, 126, 1975.

115. Huber, B., Gershon, R. K., and Cantor, H., Identification of a B-cell surface structure involved in antigen-dependent triggering: absence of this structure on B-cells from CBA/N mutant mice, *J. Exp. Med.*, 145, 10, 1977.

116. Ahmed, A., Scher, I., Sharrow, S. O., Smith, A. H., and Paul, W. E., B-lymphocyte heterogeneity: development and characterization of an alloantiserum which distinguishes B-lymphocyte differentiation alloantigens, *J. Exp. Med.*, 145, 101, 1977.

117. Subbaro, B., Ahmed, A., Paul, W. E., Scher, I., Liebermann, R. et al., Lyb-7, a new B cell alloantigen controlled by genes linked to the IgG locus, *J. Immunol.*, 122, 2279, 1979.

118. Huber, B., Antigenic marker on a functional subpopulation of B cells controlled by the I-A subregion of the H-2 complex, *Proc. Natl. Acad. Sci. U.S.A.*, 76, 3460, 1979.

119. Hodes, R. J., Hathcock, K. S., and Singer, A., Major histocompatibility complex restricted self-recognition by B cells and T cells in responses to TNP-Ficoll, *Immunol. Rev.*, 69, 25, 1982.

120. Singer, A., Morrissey, P. J., Hathcock, K. S., Ahmed, A., Scher, I., and Hodes, R. J., Role of the major histocompatibility complex in T cell activation of B cell subpopulations. Lyb-5$^+$ and Lyb-5$^-$ B cell subpopulations differ in their requirement for major histocompatibility complex-restricted T cell recognition, *J. Exp. Med.*, 154, 501, 1981.

121. Teale, J. M., Pike, B. L., Craig, J., Nossal, G. J. V., and McKenzie, I. F. C., Clonal analysis of B-lymphocyte subpopulations separated on the basis of Lyb surface antigens, *Cell. Immunol.*, 55, 272, 1980.

122. Schreier, M. H., Andersson, J., Lernhardt, W., and Melchers, F., Antigen-specific T-helper cells stimulate H-2 compatible and H-2 incompatible B-cell blasts polyclonally, *J. Exp. Med.*, 151, 194, 1980.

123. Julius, M. H., Chiller, J. M., and Sidman, C. L., Major histocompatibility complex-restricted cellular interactions determining B cell activation, *Eur. J. Immunol.*, 12, 627, 1982.

124. Hodes, R. J., Kimoto, M., Hathcock, K. S., Fathman, C. G., and Singer, A., Functional helper activity of monoclonal T cell populations: antigen-specific and H-2 restricted cloned T cells provide help for *in vitro* antibody responses to trinitrophenylpoly(LTyr,Glu)-poly(DLAla)-poly(LLys), *Proc. Natl. Acad. Sci. U.S.A.*, 78, 6431, 1981.

125. Augustin, A. A., Julius, M. H., and Cosenza, H., Antigen-specific stimulation and trans-stimulation of T cells in long-term culture, *Eur. J. Immunol.*, 9, 665, 1979.

126. Nossal, G. J. V. and Pike, B. L., Clonal anergy: persistence in tolerant mice of antigen-binding B lymphocytes incapable of responding to antigen or mitogen, *Proc. Natl. Acad. Sci. U.S.A.*, 77, 1602, 1980.

127. Pike, B. L., Kay, T. W., and Nossal, G. J. W., Relative sensitivity of fetal and newborn mice to induction of hapten-specific B cell tolerance, *J. Exp. Med.*, 152, 1407, 1980.

128. Jaworski, M. A., Shiozawa, C., and Diener, E., Triggering of affinity enriched B cells. Analysis of B cell stimulation by antigen-specific helper factor or lipopolysaccharide. I. Dissection into proliferative and differentiative signals, *J. Exp. Med.*, 155, 248, 1982.

129. Shiozawa, C., Longenecker, M. B., and Diener, E., In vitro cooperation of antigen-specific T cell-derived helper factor, B cells and adherent cells or their secretory product in a primary IgM response to chicken MHC antigens, *J. Immunol.*, 125, 68, 1982.

130. Swain, S. L., Howard, M., Kappler, J., Marrack, P., Watson, J., Booth, R., Wetzel, G. D., and Dutton, R. W., Evidence for two distinct classes of murine B cell growth factors with activities in different functional assays, *J. Exp. Med.*, 158, 822, 1983.

131. Dutton, R. W., Wetzel, G. D., and Swain, S. L., Partial purification and characterization of a BCGFII from EL4 culture supernatants, *J. Immunol.*, 132, 2451, 1984.

132. Nakanishi, K., Howard, M., Muraguchi, A., Farrar, J., Takatsu, K., Hamaoka, T., and Paul, W. E., Soluble factors involved in B cell differentiation: identification of two distinct T cell-replacing factors (TRF), *J. Immunol.*, 130, 2219, 1983.

133. Muraguchi, A., Kishimoto, T., and Miki, Y., T cell replacing-factor (TRF)-induced IgG secretion in a human B blastoid cell line and demonstration of acceptors for TRF, *J. Immunol.*, 127, 412, 1981.

134. Keller, D. M., Swierkosz, J. E., Marrack, P., and Kappler, J. W., Two types of functionally distinct, synergizing helper T cells, *J. Immunol.*, 124, 1350, 1980.

135. Zubler, R. H. and Glasebrook, A. L., Requirement for three signals in "T-independent" (lipopolysaccharide induced) as well as T dependent B-cell responses, *J. Exp. Med.*, 155, 666, 1982.

Chapter 6

B CELL TUMORS AS MODELS OF B LYMPHOCYTE DIFFERENTIATION

Larry W. Arnold, Paula M. Lutz, Christopher A. Pennell, Gail A. Bishop,
Ronald B. Corley, Geoffrey Haughton, and Nicola J. LoCascio

TABLE OF CONTENTS

I. INTRODUCTION

Lymphocytes are the cells responsible for elaborating the recognition elements of the immune response; those molecules which make it possible to distinguish between different antigens. Even extremely similar antigens can be distinguished by this system; self can be discriminated from nonself; a huge array of viruses, bacteria, eukaryotic parasites and their products, and even closely related synthetic molecules can be recognized and bound specifically by these elements. It seems likely that over 10^6 different antigenic structures can be discriminated by the recognition elements capable of being produced by the lymphocytes of a single vertebrate animal. Furthermore, two distinct sets of lymphocytes, T cells and B cells, produce different types of recognition elements which are chemically dissimilar, which serve different physiological functions, and which are the products of entirely separate sets of genes.[1,2] Those produced by B cells are ultimately secreted into the circulation and onto mucosal surfaces as immunoglobulin molecules, where they mediate a variety of effects after combination with antigen, whereas those produced by T cells remain associated with the synthesizing cell, where they function as receptors to trigger cell-mediated functions. The differentiation pathways which lead to the development of this capability for exquisitely specific antigen recognition and the means by which it is controlled are of considerable interest both to those who study the immune response itself and to those for whom it represents an accessible model for studying the fundamental mechanisms of cellular differentiation and gene control.

The early studies of lymphocyte differentiation were mainly focused on T cells, largely because T cells at different stages of differentiation could be obtained relatively easily, polymorphic differentiation antigens could be detected, and transplantable clones of T cells in the form of T lymphomas were readily obtained.[3-5] More recently, as similar approaches have become available, there has been a considerable increase in studies of the differentiation pathways, activation factors and signals, and regulatory controls of B lymphocytes.

One useful approach to studying any complex, multicomponent system is to separate the components, study each in isolation, and then attempt to synthesize an understanding of the system as a whole, from knowledge of its individual components. Such a synthesis can then be used to design experiments using normal mixed cell populations to test the validity of the assembled model. This approach would be useful for dissecting the ontogeny and differentiation controls of B lymphocytes. Ideally, it would be best to conduct such a study using normal B cells. However, there are severe practical limitations to this approach.

Although single cell isolation techniques can be used to study individual cells, the heterogeneity of B cells in normal tissues with respect to both antigen specificity and differentiative stage, as well as the inevitable admixture of other types of cells, have made it nearly impossible to study truly homogeneous populations of normal B cells. However, in recent years, considerable amounts of useful information have been obtained from study of neoplastic clones of B cells and these studies form the subject matter of this review.

II. B LYMPHOCYTE DIFFERENTIATION

There are several steps in B cell ontogeny that are characterized by changes in the arrangement and expression of immunoglobulin genes. These steps are outlined in Figure 1. All the cells comprising the hematopoietic cell lineages appear to derive from a common, multipotential stem cell. Evidence for this derives initially from studies by Till and McCulloch,[6] in which bone marrow cells from normal mouse donors were

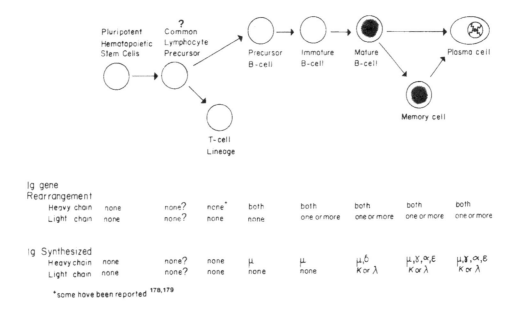

	Pluripotent Hematopoietic Stem Cells	? Common Lymphocyte Precursor	Precursor B-cell	Immature B-cell	Mature B-cell		Plasma cell

Ig gene
Rearrangement

Heavy chain	none	none?	none[*]	both	both	both	both	both
Light chain	none	none?	none	none	one or more	one or more	one or more	one or more

Ig Synthesized

Heavy chain	none	none?	none	μ	μ	μ,δ	μ,γ,α,ε	μ,γ,α,ε
Light chain	none	none?	none	none	none	κ or λ	κ or λ	κ or λ

[*]some have been reported [178,179]

FIGURE 1. Differentiation stages of B cell ontogeny. The steps in B cell ontogeny are characterized by genomic changes in the arrangement and expression of immunoglobulin genes.

injected in limiting numbers into lethally irradiated histocompatible recipients. Colonies of blood cells of donor origin were formed in the recipients' spleens and these colonies could be shown to contain differentiating erythrocytes, granulocytes, and megakaryocytes. Each of these colonies was shown, using chromosome markers, to be derived from a single colony forming unit-spleen stem cell (CFU-s stem cell). Discrete lymphocytic colonies were not seen, but when a single CFU-s colony derived from a single precursor was dissected free and injected into other lethally irradiated recipients, all lineages of myeloid, as well as lymphoid, cells developed and were derived from the chromosomally marked donor cells.[7] It is currently unknown whether the common hematopoietic stem cell first gives rise to a replicating lymphoid stem cell having the capability of differentiating along either B or T cell lineages, or whether the T and B cell lineages are derived directly and independently from the hematopoietic stem cell. However, a split between the T and B cell lineages does occur early in ontogeny and subsequent T cell differentiation proceeds within the thymus.[7] In adult mammals, the production of early cells of the B cell lineage occurs primarily in the bone marrow.[9]

The differentiation of cells within the B cell lineage along the pathway toward antibody secretion has a number of recognizable steps. These steps include both activation and differentiation events. We shall choose to make a distinction between "activation" and "differentiation". The word "activation" will be used for potentially reversible phenotypic changes as opposed to "differentiation", which involves permanent genetic change resulting in phenotypic alteration. Activation in response to environmental influences may be necessary to prepare or allow certain cells to initiate differentiation. The complex stages associated with B cell development are best and most easily simplified to a framework based on permanent genetic changes. In the B cell lineage these changes are associated with rearrangements in the genes for both the heavy and light chains of immunoglobulin molecules.[1,10,11] The CFU-s and common lymphocyte precursor (if it exists) probably do not express immunoglobulin gene rearrangements, since many of the other cell types (monocytes, granulocytes, T lymphocytes) derived from these precursor cells do not usually have any such rearrangements.[10,12,22] The first recognizable cell of the B cell lineage has undergone immunoglobulin gene rearrangements that involve the heavy-chain loci only.[12,13] The dispersed arrangement of DNA segments coding for the heavy-chain variable region, as found in the germline, has been altered to bring a single V segment into contiguity with a single D and a single J segment. This forms a functional variable region gene which can now be transcribed, along with the nearby Cμ gene, to yield mRNA coding for the entire heavy chain of IgM. Light-chain gene rearrangement and synthesis and δ heavy-chain synthesis follow to produce the mature "resting" B cell. Following a number of activation steps the mature resting B cell differentiates to a plasma cell secreting immunoglobulin or to a memory cell.[14] The plasma cell or memory B cell may undergo further gene rearrangements, by which the assembled VDJ sequence is shifted from proximity with the Cμ gene to bring it close to one of the other heavy-chain constant region genes, encoding the other (γ, α, ε) isotypes of immunoglobulin.[1,11] This last process is usually called "class switching". Accompanying these stages of differentiation is a variety of changes in surface membrane antigen expression.[12,15,21] Some of these markers may define functionally distinct stages of B cell maturation or indeed distinct branched lineages as recently suggested.[10]

III. APPROACHES TO STUDY B CELL DIFFERENTIATION

The isolation of cells at any one of the distinct stages of B cell differentiation (and there may be additional undefined stages or branches) is difficult. In an attempt to achieve pure cell populations, highly selective procedures must be employed which may result in unknown and undesirable selection or alteration of the properties of the purified cells. Attempts to isolate pure populations of cells are based on two requirements. The population to be purified must be identified by some marker which will remain stable throughout the purification. Additionally, the cell type must be present in sufficient quantity to permit isolation of useful numbers of cells. Even when substantial numbers of cells can be isolated for cellular studies, these may provide insufficient material for biochemical, immunochemical, or molecular biological analyses. Unwitting contamination with small numbers of physiologically important cells is a further problem with the use of physically separated normal cell populations in these types of experiments.

The difficulties associated with physical separation of normal lymphoid cells, homogeneous with respect to stage of differentiation, are compounded by considerations of antigen specificity. Not only are these cells heterogeneous with respect to stage of ontogeny, but also in terms of their antigen receptors. Immunocompetent lymphocytes express cell surface receptors for antigen. These receptors are immunoglobulin molecules in the case of B lymphocytes, and an immunoglobulin-like molecule encoded by a distinct set of genes in the case of T cells.[2,22] The binding of antigen to these receptors plays a central role in the activation and differentiation of these cells into functional effectors (*vide infra*). The total antigen repertoire of the immune system is estimated at between 10^6 and 10^8 specificities. This poses a virtually insuperable barrier to the isolation, by physical methods, from normal lymphoid tissues, of substantial numbers of lymphocytes, homogeneous with respect to both differentiative state and antigen specificity. In order to reduce this complexity to more nearly manageable proportions, polyclonal activators such as bacterial lipopolysaccharide (LPS), dextran sulfate (DXS), and antiimmunoglobulin antibody rather than specific antigen have often been used to induce differentiation of sufficiently large numbers of B cells to permit study.[23-28] The use of antiimmunoglobulin to induce B cell activation/differentiation is discussed in detail in another chapter of this publication. While these reagents have been useful for investigating a number of aspects of B lymphocyte biology, they are not physiologically relevant. Attempts to study activation and differentiation of B cells by physiologically relevant (i.e., antigen-driven) pathways are extremely difficult. Consequently, the relationship between the requirements for and processes of differentiation induced by polyclonal activators and those resulting from interaction with specifically bound antigen has not yet been established in any critical way.

What are the solutions to these many problems and limitations? The use of cloned and cultured normal cell populations would solve the problems of cell purity and number, and such approaches have been used. In recent years, attempts to develop techniques for the culture of normal lymphocytes have been most successful for T cells.[5,29,30] The culture of B lymphocytes and, in particular, cloned, antigen-specific B cells has met with limited success.[31,32] These techniques are discussed in detail in an earlier chapter of this volume. Despite remarkable advances, cloned, continuous cultures of normal B cells at all phases of ontogeny have not been established.

Another approach to the analysis of B lymphocyte ontogeny and differentiation is to use immortal, clonal cell lines including lymphomas and plasmacytomas. We will discuss, in this paper, how several different malignant B cell lines have been used to define the genetic and phenotypic characteristics of the various stages of B cell differentiation. The emphasis will be on murine systems, but considerable data are becoming available from studies of human tumors and these will be discussed when appropriate. Also, we will show how these types of cells have been used to dissect the mechanisms by which B cell activation and differentiation are regulated.

Lymphoid malignancies are usually examples of unlimited clonal proliferation. The majority of the cells within a single tumor are "arrested" at a given maturational level and, therefore, represent a homogeneous population. However, as we will discuss in this review, many of these tumor cells retain at least some ability to differentiate. A great deal of evidence from many laboratories (vide infra) has validated the use of malignant cell lines as models of normal differentiation. This is particularly true when tumors are used within a relatively small number of cell generations from the primary malignancy. Nevertheless, the potential of tumor cells for aberrant responses can never be overlooked and the data obtained from these model systems should be used to guide investigations with normal cells whenever possible. It has become apparent that some of these cell lines exhibit changes in functional properties with continued passage.[33] Therefore, one must be careful interpreting the data. Some of the discrepancies found within data using the same tumor may well be due to differences between tumor sublines. We will begin our discussion with examples of how B cell tumors have been used to provide information about the various stages of the B cell differentiation pathway and conclude with studies of how these tumors may be used to investigate the regulatory mechanisms which trigger progression from one differentiation stage to another.

IV. PLASMACYTOMAS — STUDY OF IMMUNOGLOBULIN STRUCTURE AND GENETICS

Historically, tumors of plasma cells, the terminally differentiated state of the B lymphocyte lineage, were the first to be studied. Plasma cells have well-developed synthetic and secretory machinery for the production of immunoglobulins. Each plasma cell normally secretes high levels of a single, unique immunoglobulin.[34] Tumors of these cells thus provided not only a large population of homogeneous cells for cellular study, but also large amounts of chemically pure antibody for molecular study. Plasmacytomas have proven extremely useful in studying immunoglobulin structure, genetics, and plasma cell biology.[35] Plasmacytomas may be readily induced in two strains of mice, BALB/c and NZB, by the simple injection of mineral oil or several of its active components, pristane being one example.[35-37] By deriving congenic strains of mice a variety of allelic forms of the heavy-chain immunoglobulin genes from mice has been placed on the BALB/c background to produce plasmacytomas secreting a variety of immunoglobulin allotypes. These tumors are characterized by a predominance of IgA-producing lines.[38] Of those cell lines for which antigen-binding specificity is known, a majority are specific for antigens expressed on microorganisms.[37] These data, in conjunction with sequence data demonstrating differences in plasmacytoma immunoglobulins from NZB and BALB/c mice, suggest that the tumors are not examples of random malignant transformation affecting the total B cell repertoire, but are derived from a limited subpopulation of B cells.[39,40]

Plasmacytomas and their immunoglobulin products were the first useful experimental systems for study of immunoglobulin biosynthesis and immunoglobulin structure.[41,42] More recently, these cells have provided materials for dissecting the complex molecular genetics of immunoglobulin genes.[1,11] The homogeneous heavy- and light-chain mRNA required to produce hybridization probes for Southern and Northern analyses have almost exclusively been provided by plasmacytomas. The detailed analyses of immunoglobulin gene organization and, in particular, rearrangements of the heavy- and light chain genes have provided many answers to fundamental questions in immunology. These subjects have been extensively reviewed in this volume and elsewhere and will not be discussed in detail here.[1,11,41-43]

V. DIFFERENTIATION ANTIGENS OF MALIGNANT B CELLS

A particularly useful application of malignant B cells is the identification and characterization of cell membrane antigens associated with B cell development and the definition of cell subsets within the differentiation pathway. Because the tumors mostly represent expanded clones of cells, each displaying an arrested stage of phenotypic development,[44] they can be useful to define those phenotypes which may be expressed only transiently or only by relatively rare cells in normal lymphoid tissue. Malignant cell lines, therefore, provide homogeneous proteins, DNA, mRNA, and other cell constituents in amounts suitable for biochemical, immunological, and genetic analysis. The advent of monoclonal antibody technology has made possible the preparation of specific, pure antibodies following immunization of animals with complex antigen mixtures. Malignant cell lines have been useful both as homogeneous sources of immunogen and for characterizing antibodies raised against heterogeneous normal cells, with respect to the cell lineage and the stage of differentiation at which the corresponding antigen is expressed. The following is a description of a number of examples in which malignant cell lines have proven useful.

A. Use of Malignant B Cells to Prepare and Define Monoclonal Antibodies

Several monoclonal antibodies specific for differentiation antigens of the B cell lineage have been produced by immunizing either mice or rats with pre-B cell tumors. Several different tumors have been used. Some (e.g., ABE-8 and RAW112) arose following transformation of B cells by the Abelson murine leukemia virus. This virus has the unique property of transforming immature lymphocytes of the B cell lineage.[45,46] Most of these cell lines lack surface immunoglobulin, but have cytoplasmic heavy chain. Another tumor often used as an example of a pre-B cell is called 70Z/3. This tumor arose in a (C57BL/6 × DBA/2)F₁ mouse following thymectomy and exposure to methyl nitrosourea.[47] 70Z/3 expresses cytoplasmic immunoglobulin heavy chains without light chains.

Dessner and Loken immunized rats with 70Z/3 tumor cells and identified monoclonal antibodies reactive with ABE-8 and 70Z/3 but not with a myeloma line.[48] Subsequent analysis showed one of these, DNL1.9, to react with plaque-forming cells (PFC), as well as membrane immunoglobulin-positive and -negative B cells. There is no evidence for the presence of the antigen detected by this antibody on T cells or stem cells (CFU-s, CFU-c, BFU-e). The 70Z/3 tumor has also been used by Hammerling and co-workers to produce monoclonal antibodies reactive with the Ly-m19 and Ly-m20 antigens.[49,50] The Ly-m19 antigen is carried by both T and B cells, while the Ly-m20 antigen is found on B but not T lymphocytes. McKearn et al.[21] used both the ABE-8 and 70Z/3 tumors to immunize rats and produce monoclonal antibodies reactive with immature B cells. These monoclonal antibodies, AA4.1 and GF1.2, detect antigens present only on early B cells. In addition, their studies suggest that the antigen detected by antibody AA4.1 appears prior to the antigen detected by GF1.2. Coffman and Weissman[51,52] produced a series of monoclonal antibodies by immunizing rats with the Abelson virus transformed cell line RAW112. One of the antibodies, RA3-2C2, stains all B lymphocyte lineage cells, including membrane immunoglobulin-negative pre-B cells, and also most plasma cells. This antibody reacts with the B220 antigen, the B cell-specific form of the T200 glycoprotein family.

Other stages of B cell tumors have also been used to prepare antibodies. The B cell lymphoma, I.29,[53] was used to immunize C3H.I-H-2^j mice to produce antisera detecting the alloantigen Lyb-2.[54] The antigen is present only on B lymphocytes in spleen, lymph node, and bone marrow. This, as well as the finding that the 70Z/3 Abelson tumor is Lyb-2 positive (whereas plaque-forming cells are negative), suggests that Lyb-2 is a marker of early B cells.[55,56] Plasmacytomas have also been used to raise antibodies directed against B cell differentiation antigens, one of which defines the ThB antigen. This antigen, which is expressed on murine B lymphocytes and a subpopulation of thymocytes, was originally defined by a rabbit antiserum raised against the BALB/c myeloma, MOPC-104E.[57] More recently, a monoclonal antibody to this antigen has been produced by immunizing rats with MOPC-104E.[58] MOPC-70A, another mineral oil-induced plasmacytoma of BALB/c mice, has been used to prepare both an antiserum[59] and a monoclonal antibody[60] which detect antigens (PC.1 and PC.2, respectively) expressed only on terminally differentiated B cells (plasma cells).

In addition to serving as sources of homogeneous immunogen, clonal malignant cells bearing constant arrays of differentiation antigens which correspond to known stages of normal lymphopoiesis have been used as reference materials for characterizing antibodies, which react with small subpopulations of normal cells. Clonal tumor cells also provide homogeneous sources of cells with which to study fluctuations in antigen expression as a function of cell cycle or culture conditions, as well as providing examples of normal, but previously undetected, small populations of cells. The use of malignant clones of cells as reference materials for the study of normal differentiation antigens depends upon the availability of a large library of tumors representative of many differentiation stages, which have been assigned to each stage using as many criteria as possible. A pioneer in this effort has been Noel Warner, who, with his colleagues, began in the 1970s to collect such a library of murine tumors. Using these tumors, together with flow cytometry and both conventional antisera and monoclonal antibodies, they assembled a large data base with which to begin to classify their tumors according to the emerging schemes for hematopoietic development.[61-63] A corresponding body of work on human B lymphoid tumors has been performed by Godal and co-workers.[64-68] These libraries of tumors can be used extensively to analyze monoclonal antibodies, changes in the expression of cell surface protein antigens associated with differentiation or cell cycle, and to define previously unrecognized subpopulations of cells as exemplified by the following.

Ledbetter and Herzenberg[69] produced a series of monoclonal antibodies reactive with murine lymphocytes. Many of these reacted with antigen previously defined by conventional antisera. However, several reacted with previously undetected cell-surface protein antigens. Lanier et al.[33] attempted to define the specificities of these antibodies using a large number of tumor lines which had been characterized for a number of antigen markers. Two of these antibodies were of particular interest. Monoclonal antibody 30-E2 was found to react with approximately half of the B cell lymphomas and several myeloid tumors, but not T cell lymphomas and plasmacytomas. This antibody has been used recently to define a subpopulation of normal B cells.[19] Antibody 30-H11, which previously was thought to react only with T lymphocytes, was found to detect an antigen on several B cell lymphomas and plasmacytomas. An explanation for this may be that the 30-H11 antigen is expressed on a very small subpopulation of normal B cells or may be present in a lower quantity on normal B cells than malignant cells, making detection difficult. Many of the tumors also exhibited marked intratumor heterogeneity for the expression of a number of antigens. One explanation for part of this heterogeneity may be quantitative differences in antigen expression related to cell cycle. Lanier and Warner[70] investigated this possibility using the WEHI-231 tumor. This is a B cell lymphoma induced with mineral oil in a (BALB/c × NZB)F$_1$ mouse.[71] Only 20 to 30% of the cells of this tumor bear Ia.7 (I-E) and Ia.8 (I-A) antigens. The data suggest that the levels of the antigens change based on cell cycle; Ia.7-positive cells are enriched in the G0/G1 phase and Ia.8-positive cells are enriched in the G2/M phase.

B. The Ly-1 B Cell Subpopulation

One of the most interesting findings to emerge from the study of B cell lymphomas in recent years has been the discovery of previously unrecognized subpopulations of B cells. Initial analyses had demonstrated that T lymphocytes were distinguishable from B lymphocytes by reactivity with a number of alloantisera/monoclonal antibodies. Among those antigens thought to reside exclusively on T cells were the Thy-1 and Lyt-1, 2, 3 antigens.[72] Lyt-1 had been thought to be expressed on immature T cells and on T helper cells, but to be absent from T cytotoxic and suppressor cells.[72] Later studies, however, demonstrated that the Lyt-1 antigen was present on all T cells, albeit at lower concentration on mature cytotoxic and suppressor cells,[73] and that a population of Thy-1-negative cells were also Lyt-1 positive.[74] During an analysis of two series of B cell lymphomas, the WEHI and CH series, for a number of cell surface antigens, four tumors were found to express Lyt-1.[75,75] Additional studies demonstrated conclusively that these tumors were, indeed, B cells and were synthesizing the Lyt-1 glycoprotein. CH5 tumor cells which express mIg (μ,κ) and appropriate Ia antigens[77] were stripped of cell surface proteins with trypsin and cultured overnight. Re-analysis for the Lyt-1 antigen showed it to be present, i.e., regenerated by the cells. None of the B cell tumors expressing Lyt-1 could be shown to also express the Thy-1 antigen. Some human B lymphocyte tumors were also reported to express the Leu-1 antigen, the human counterpart of the murine Lyt-1.[78-80]

Possible explanations for these findings included: (1) B cell lymphomas express T cell markers aberrantly and inappropriately, or (2) a subset of normal B cells exists which expresses Lyt-1. Following these observations, a renewed and successful attempt was made to identify the Lyt-1 (now called Ly-1)-positive normal B cells.[81,82] The normal Ly-1 B subpopulation represents about 2% of normal spleen cells and these cells bear high levels of mIgM and low levels of mIgD as revealed by flow cytometric analysis.[83] Interestingly, these cells account for 5 to 10% of the spleen cells from NZB mice, a strain characterized by hyperactive B cells and high levels of serum IgM and autoantibody. The Ly-1 B cells appear to account for the spontaneous IgM secretion seen in NZB spleen cells[83] and to be responsible for the autoantibodies produced in this strain. These observations have prompted the suggestion that the Ly-1 B-cells are a functionally distinct subpopulation of B cells.[84,85]

The recent report of an AKR B cell lymphoma which expresses both Lyt-1 and Lyt-

2 raises the question of whether there is also a normal subpopulation of B cells which has this phenotype.[86] The AKR225 tumor is a spontaneous lymphoma which arose in an old AKR mouse. Analysis of its membrane antigens shows that >80% of the cells bear Ig ($\mu,\varkappa$), IA and IE, ThB, E2, DNL1.9, and B220, in addition to Lyt-1 and Lyt-2. The cells do not bear the T cell restricted antigen, Thy-1. Other investigators have previously reported AKR lymphomas which express both T and B cell markers.[87] However, a more extensive study showed that these tumors consisted of two cell types, which could be separated using flow cytometry sorting and which could be grown independently.[87] This explanation is not likely in the case of the AKR225 tumor, since >80% of the tumor cells were positive for both Lyt-1 and 2, but has not been formally excluded. The recent report of a B cell lymphoma in the CH series, which bears Lyt-2, adds to the credibility of the speculation that there may be a population of normal B cells with this phenotype.[89] To date, we are unaware of any report of a normal subpopulation of B cells which have been shown to bear both Lyt-1 and 2.

VI. STUDIES OF INDUCED DIFFERENTIATION OF MALIGNANT B CELLS

Perhaps one of the most useful applications of malignant B cells is their use as a source of homogeneous cells to analyze the complex interactions and mechanisms responsible for the differentiation of B cells. An early concept of malignant lymphocytes was that the cells were aberrant forms irretrievably "frozen" at a particular stage of growth. This, in fact, is not always so. In the late 1970s, reports appeared which showed that a number of malignant B cell lines could be induced to differentiate.[89-92] Since that time, considerable effort has been devoted to studying the differentiation of B cell tumors. A discussion of a number of experimental tumor model systems employed to study B cell differentiation is provided.

A. Pre-B Cell Tumors

The earliest recognized stage of B cell differentiation for which corresponding tumors have been identified is the so-called pre-B cell stage, i.e., the stage at which B cells show IgH gene rearrangement and cytoplasmic heavy chains are present, but not surface immunoglobulin. A number of tumors at this stage of differentiation exists, the majority of them having arisen following transformation by the Abelson murine leukemia virus (A-MuLV).[45,94] This virus, a defective leukemia virus which can only propagate in the presence of a helper virus, infects cells of the B lymphocyte lineage with a preference for cells at very early stages of differentiation. Two levels of pre-B cell differentiation have been distinguished using these cells as well as the 70Z/3 tumor (vide infra). The earliest pre-B cells appear to have immunoglobulin heavy-chain gene rearrangements only, while a later stage appears to have both heavy- and light-chain genes rearranged in the absence of light-chain synthesis. These findings are inferred from early studies which showed that while replication of most Abelson lines would be stimulated or inhibited by the polyclonal activator, LPS, only a few could be induced to synthesize light chain.[94,95] Also, fusion of some Abelson pre-B cell lines with nonsecreting myeloma cells resulted in hybrids having the same heavy-chain V gene but using random light-chain V genes, suggesting that light-chain gene selection had not occurred in the Abelson tumor.[97] More recently, Coffman and Weissman[12] have described the gene rearrangements in murine pre-B cells and can distinguish the two stages described. Two other murine pre-B cell tumors have been described which have proven useful. These cell lines, 70Z/2 and 70Z/3, were induced with methyl nitrosourea.[98] Like the Abelson tumors, these cells are cytoplasmic immunoglobulin positive, mIgM negative, and Ia antigen negative. Both of these tumors may be induced with polyclonal activators to express surface IgM.[47] The 70Z/3 tumor has rearranged one kappa chain gene and the other gene is in the germline configuration.[99]

Most of the Abelson tumors and the 70Z tumors are responsive to polyclonal B cell activators. Ralph et al.[100] found that a number of pre-B cell tumors could be growth inhibited by exposure to either dextran sulfate, LPS, or PPD (purified protein derivative). A number of these lines also expressed surface immunoglobulin following exposure to these agents.[47,101] These findings are consistent with the observation that once the commitment to differentiation to a mature B cell is made, the pre-B cell rapidly leaves the cell cycle.[102]

Much of the work on the differentiation of pre-B tumors has been done with the 70Z/3 tumor. As stated above, the 70Z/3 tumor responds to exposure to LPS by synthesis of κ light chains and expression of mIgM. Although one kappa chain gene is rearranged in this tumor,[99] very little kappa mRNA can be detected.[102,103] Therefore, the exposure to LPS must induce a transcriptional change in the cells. Interestingly, Parslow and Granner[104] found that following exposure to LPS, a discrete region linked to the coding sequence for the kappa chain constant region becomes DNase sensitive. This LPS-inducible hypersensitive site may play a role in the transcriptional response to the mitogen. Unlike LPS, dextran sulfate induces 70Z/3 to express surface μ chains without light chains.[103] The authors speculate that perhaps surface heavy-chain expression without light-chain induction can occur in normal pre-B cells, but may not be a stable condition. In addition to the mitogens LPS and DxS, factors from myeloid and lymphoid cells may also activate pre-B cells to express mIgM. Examples include the activation of 70Z/3 with IL-1[105] and T cell factors.[106]

B. Mature B Cell Malignancies

In recent years, the use of more highly differentiated surface immunoglobulin-positive B cell tumors, including human malignancies, has provided a great deal of information about B cell differentiation. Chronic lymphocytic leukemia, a B cell proliferative disease, represents about 25% of all human leukemias.[90,92] However, until recently only a few B cell tumors existed in easily manipulated experimental animals, such as the mouse. Several such tumors have now been identified and used to study B cell differentiation. With a few isolated exceptions, the B cell lymphomas occur as groups of tumors related by either species or strain of animal and induction protocol, or both.

Chronic lymphocytic leukemia (CLL) in humans represents one of the largest groups of B cell malignancies. Most patients with CLL are older than 50 years[112] and the disease does not rapidly cause severe or fatal complications, hence the term chronic. The most significant manifestations, in addition to the leukemia, are the marked infiltration by leukemic cells of lymphoid tissues such as lymph nodes and spleen. The circulating leukemic B cells have a morphology resembling small quiescent lymphocytes and the majority bear both membrane IgM and IgD,[107] HLA-DR antigens,[108] and Fc and C3 receptors.[109] The IgM and IgD molecules have the same idiotype and, therefore, use the same V_H gene.[110,111] A small group of CLL cases has been reported which exhibits a monoclonal serum IgM with the same idiotype as the leukemia cells.[111,113,114] This implies that at least some of the leukemic cells retain the capacity for differentiation to plasma cells. Other human B cell malignancies exist, but have generally been studied less than CLL.[115]

A number of B cell malignancies has been identified in mice and the derivation of some of these are described here. Warner and colleagues[75,116] have described a series of tumors which arose in New Zealand mice, F1 hybrids between these mice and other strains, or in BALB/c mice. These tumors (the WEHI and UNM series) arose spontaneously or following injection of mineral oil. All are surface immunoglobulin positive and appear to represent stages of differentiation between mature B cells and plasma cells.[61,75] In addition to mIg, the cells express the appropriate IA and IE antigens and lack Thy-1 and Lyt-2. Some are Ly-1 positive as described above.

Kim et al.[117] have described a group of BALB/c B cell lymphomas which arose spontaneously or after injection of methyl-1-nitrosourea. All six of these tumors were mIg positive and lacked the Thy-1 antigen. They varied in membrane Ia antigens (five were positive) and all were Fc receptor positive, C3 receptor negative.

The BCL$_1$ lymphoma, which arose spontaneously in a BALB/c mouse, has been widely studied.[118,119] It causes marked splenomegaly, with a leukemic phase occurring in the later stages of the disease. The cells bear membrane μ and δ heavy chains and λ light chains, Fc receptors, and IE molecules. However, IA molecules and complement receptors have been reported to be absent.[121] Both in vivo and in vitro lines of BCL$_1$ exist. The in vitro line is nearly tetraploid (modal chromosome number of 60) in comparison to the in vivo line, which is near diploid (modal chromosome number of 35).[119]

A particularly curious group of lymphoid tumors which arose in B10.H-2^{a}H-4^{b}p/ Wts mice has recently been described.[75,121,122] These tumors occurred, for the most part, in mice that had been adoptively hyperimmunized against sheep erythrocytes. The series of lymphoid tumors that arose in the adoptively immunized mice includes both B cell and T cell lymphomas. A group of the T cell lymphomas, part of the UNC series, has been studied by Corley and associates[123] and shown to be diverse with respect to ability to produce specific single lymphokines. However, the B cell lymphomas, the CH series, have been studied in greater detail. Twenty-seven CH series B cell lymphomas have been described, all of which express cell-surface IgM, as well as K, D, IA, and IE molecules appropriate to their H-2^a haplotype. Only one of the tumors, CH15, bears mIgD in addition to mIgM, and none has yet been described as bearing any other immunoglobulin isotype. The Thy-1 antigen is not found on the cells of any of these tumors, but all of the tumors tested express detectable levels of Ly-1. Receptors for the Fc of immunoglobulin are present on most, but not all, of the tumors, whereas C3 receptors are absent from most. None of the tumors harvested from syngeneic hosts during the early or intermediate stages of growth in vivo contains more than 1% of Ig secreting cells detectable by the reverse plaque assay,[77,124] although in the case of CH12, this frequency increases to about 3% during the terminal stages of tumor growth.[124] The antigen specificity of the mIg of 14 of the 27 tumors is known.[89,124] Nine of the tumors have specificity for bromelain-treated mouse erythrocytes. Of these nine, six also react with sheep and chicken erythrocytes and one reacts with all three erythrocytes as well as Escherichia coli. Five other tumors also react with E. coli, but not with any of the erythrocytes. The idiotypes of the surface immunoglobulin of the 27 CH lymphomas have been studied by means of xenoantisera and have been found to display extensive cross reactivity. Of the 27 tumors, 22 are related by either antigen or idiotype cross reactivity, or both.[89,125] These findings imply that the B lymphomas of 2^{a}4^b mice are derived mainly from a subset of normal B cells, characterized by the idiotype or antigen specificity of the Ig born on their surfaces. The factors controlling lymphomagenesis are not fully understood, but a role for genes associated with H-2^a and H-4^b is clear, with H-2 exerting the major influence.[180] It has recently been found that a gene within the H-2 complex regulates expression of the CH12 idiotype in normal splenic B cells.[181]

1. Antigen Presentation by Neoplastic B Cells

These various neoplastic B cell lines have been useful in elucidating and dissecting a number of functional capabilities, differentiation pathways, and signals for B cell activation. An interesting capability of B cells recently illuminated is that of antigen presentation. The subject is discussed in detail in a preceding chapter of this book. B lymphocytes recognize antigen directly via interaction of their mIg receptor with epitopes on the antigen molecule. T cells, on the other hand, recognize a complex between antigen and the class II molecules of the major histocompatibility complex; this complex is not yet fully characterized.[126] The ability of cells such as macrophages, Kupffer cells, Langerhans' cells, and dendritic cells to present antigen to T cells has been recognized.[127-129] Recent evidence,[130] however, demonstrates that B cells can also present antigen and B cell tumor lines have provided a means of studying this phenomenon. Walker et al.[131] showed that several B cell lymphoma lines can present soluble, but not particulate, antigens to T cells as detected by T cell proliferation. Glimcher et al.[132] reported similar results using a similar assay system, but different B cell tumor lines. These same authors later showed that hybrids between one of these lymphoma lines and normal splenic B cells would express I region-encoded molecules of both parents, thereby permitting the generation of antigen-presenting cells with Ia antigens from a variety of H-2 haplotypes.[132] These cell lines, because of their homogeneity and availiability in large numbers, may prove useful in understanding the biochemical nature of antigen presentation.

2. Polyclonal Activation of Malignant B Cells

Study of malignant B cell lines has provided information about the initiation and control of differentiation of B cells. In the intact animal, normal B lymphocytes are activated by exposure to antigen along with appropriate T cell help. Since only those cells recognizing the antigen are stimulated, the number of B cells in an unprimed animal responding to a given antigen is small, which makes study difficult. To overcome this difficulty and in order to provide sufficient cells for study, much use has been made of polyclonal activators such as bacterial lipopolysaccharide (LPS) or anti-Ig antibody reactive with constant region determinants on the mIg receptor. While an argument can be made that anti-Ig mimics the action of antigen and by-passes the constraints of specificity since it binds to the same receptor molecule, it is more difficult to support such an argument on behalf of LPS, which binds to a receptor different from mIg. In addition to their use with normal B cells, both of these types of polyclonal activator have been used to activate clonal B lymphoma cells, thereby reducing the complexity of the system. Finkelman et al. provide a detailed discussion of the use of anti-Ig reagent to induce B cell differentiation in another chapter of this book.

A number of tumors have been shown to exhibit activational changes following exposure to LPS. The BCL_1 tumor, both in vivo and in vitro lines, has been shown to increase or at least maintain proliferation and to increase the number of plaque-forming cells (PFCs) in response to LPS.[135,136] The WEHI-231 B cell lymphoma has been reported to secrete IgM in response to LPS.[137] Other investigators, however, have failed to see either proliferation or increases in secreted IgM in WEHI-231 or other tumors, and in some cases, inhibition of proliferation was seen.[138] These data are in agreement with those of Ralph and co-workers who showed that cell growth of several B lymphoma lines was inhibited by LPS.[99,139] The CH12 lymphoma,[140] as well as several other CH lymphomas (unpublished observations), respond to LPS with both cell proliferation and increase in secreted IgM. Some in vitro subclones of CH12, however, have lost the ability to respond to LPS, while others retain this responsiveness (unpublished observations). The discrepancy in LPS reactivity among tumors may be due in part to differences between tumor sublines or in subtle differences in the stage of differentiation. Also, since most ($\sim^2/_3$) normal B cells do not respond to LPS (perhaps some are actually inhibited), it may not be surprising that B cell tumors reflect this differing reactivity.

In addition to measuring proliferation and Ig secretion following exposure to LPS, changes in the transcription or translation of Ig genes and changes in cell surface markers other than Ig have been studied. This may help establish whether there is an ordered acquisition of differentiation markers and, if so, what that order is. The BCL_1 tumor responds to LPS with a decrease in mIgM and an increase in secreted IgM.[141] These changes are apparently independent of cell cycle and of *de novo* DNA synthesis. No discernible change in the mRNA of the two forms of Ig was seen, implying post-transcriptional control. Yuan and Tucker, however, found that following LPS stimulation, the level of mRNA for the secreted form of Ig increased, while that of the membrane form decreased slightly.[142] This suggested both transcriptional and translational regulation. Yuan[143] also found that blockade of glycosylation with tunicamycin prevented BCL_1 from expressing IgM and prevented the LPS-induced production of secreted IgM.

Lanier[138] has found changes in differentiation antigens expressed on B cell lines following LPS exposure. Although several tumors showed no significant changes, two tumors, L10A/2J and NBL-DU-B, did. Both I-A and I-E antigens were increased on L10A/2J cells and mIgM and ThB were increased on NBL cells. We have seen changes in cell surface markers on CH12 following LPS treatment (unpublished observations). Membrane IgM, as well as both I-A and I-E on CH12, is significantly decreased compared to cells cultured without LPS. The differences between tumors are probably due to their representation of different stages of differentiation. Those tumors which are relatively immature may show increased levels of mIgM and Ia, while the more mature cells may show less of these antigens following activation.

B cell lymphomas have also been influenced by antiimmunoglobulin antibodies, as well as by various T cell factors. BCL_1 has been reported to be activated by anti-μ or anti-δ.[144] The cells proliferate in response to anti-Ig, but only secrete Ig following the addition of T cell factors. Godal and colleagues have studied the effects of antiimmunoglobulin reagents on DNA synthesis in human B lymphoma cells. They found that in most cases, the anti-Ig alone could not induce DNA synthesis, but could in the presence of the tumor promoter TPA.[145,146] Anti-μ and anti-δ elicited different responses,[147] with both inducing DNA synthesis, but only anti-μ significantly increased Ig synthesis. These investigators also found differences in the activation properties of monoclonal vs. polyclonal antibodies.[148] Monoclonal antibodies could induce early activation events, but could not induce DNA synthesis as could the polyclonal antibody.

Kishimoto and co-workers have shown that human CLL cells can be induced to proliferate and secrete Ig in the presence of T cells and a T cell mitogen or with an anti-Ig and T cell factors.[149,150] The T cell factors alone or anti-Ig alone could induce neither response. Also, F(ab')$_2$ fragments of anti-Ig were as effective as whole Ig, but Fab' fragments would not work, suggesting a role for cross-linking of the mIg molecules and also indicating that the Fc portion of the molecule was not involved. The anti-Ig reagent could be directed against either constant region determinants or idiotype determinants. Separation of the T cell factors into IL-2 and TRF-containing fractions showed that anti-Ig and IL-2 could induce proliferation, but not Ig secretion; however, anti-Ig and TRF could do neither. Anti-Ig, IL-2, and TRF were required to produce Ig secretion. The IL-2 was only necessary during the first 48 hr of culture, whereas the TRF was not required until after 48 hr.

These findings differ from others, which suggest that some lymphoma lines can be induced to differentiate by T cell factors alone.[151-153] Pure et al.[151] reported that BCL_1 could secrete IgM following exposure to T cell factors and that IL-2 was not involved. The human B cell line, CESS, is also reported to secrete Ig in response to exogenous TRF. This cell line could absorb the TRF activity, suggesting the presence of TRF receptors on the cells. T cell-growth factor (IL-2) was not required. This cell line could also respond, by secretion of Ig to B cell growth factor (BCGF) and B cell differentiation factor, produced by T cell hybridomas. These disparate findings suggest that the tumors are used in different stages of differentiation and therefore vary in their ability to respond to anti-Ig and T cell factors.

In addition to providing stimulatory signals, anti-Ig also has been shown to inhibit the proliferation of some B cell tumors. Boyd and Schrader found that anti-Ig prevented cell proliferation and resulted in the death of WEHI-231 cells. Both whole anti-Ig polyclonal and monoclonal antibodies as well as F(ab')$_2$ fragments were effective, but Fab' fragments were not. Antisera against other membrane components did not produce these effects. Proliferation of the BCL$_1$ tumor is also inhibited by anti-Ig antibodies.[154,155]

3. Induced Differentiation of Antigen-Specific Malignant B Cells

Although the studies discussed above have provided considerable information about the activation of B cells to proliferate and secrete Ig by anti-Ig and T cell factors, it has never been formally shown that anti-Ig mimics the physiologic activation by antigen binding. The limitation has been the lack of a clonal homogeneous population of B cells for which the antigen specificity is known. Several such populations have now become available in the form of the CH series of B cell lymphomas described above. One of these tumors, CH12, has been analyzed in detail. CH12 bears mIg (μ,$\varkappa$) which binds to sheep red blood cells (SRBC).[156] This antibody also reacts with chicken red blood cells (ChRBC) and bromelain-treated mouse cells (BrMRBC). CH12 expresses both I-A and I-E molecules as well as Ly-1, ThB, E2, DNL1.9, RA3-2C2, and Ly-5.[156] The binding of SRBC to CH12 cells was shown to be via the mIgM, by inhibition of binding with anti-μ or specific antiidiotype antisera. These reagents, when used under conditions which allow the mIgM molecules to cap off the cell, also removed the ability to CH12 to bind SRBC. Less than 1% of the cells in the early and intermediate stages of growth in vivo secretes detectable anti-SRBC antibody. During the growth of the tumor in mice, anti-SRBC levels in the serum become detectable in the late stages of growth and increase sharply. This increase is greater than can readily be explained by cell growth since the PFC increase at a rate greater than the increase in total cell number. Growth of CH12 in vitro is also accompanied by an increase in SRBC-specific PFC. These data, taken together, indicate that CH12 is able to differentiate to an antibody-secreting cell, i.e., is not irretrievably "frozen" in its differentiation.

Later studies examined the ways in which CH12 could be induced to differentiate.[140] Culture of CH12 with LPS induced both proliferation and differentiation. About 10% of the cells were able to differentiate to PFC. The presence of specific antigen, SRBC, did not affect the LPS dose-response curve, indicating that mIg binding did not provide any synergistic signal for differentiation induced by LPS. The cells of CH12 could also be induced to secrete hemolytic antibody when cultured with SRBC-primed or ChRBC-primed T cells, but not T cells primed with other antigens. The presence of SRBC was an absolute requirement, and no bystander effect could be demonstrated by using T cells primed with other antigens, even when the T cell-priming antigen as well as SRBC was cultured with CH12.[140] The T cells responsible for the help were shown to be Lyt-1$^+$, 2$^-$ helper cells. In contrast to some other tumors (*vide supra*), CH12 could not be induced to differentiate by T cell-derived soluble factors, even in the presence of SRBC. This indicated that linked recognition of antigen, by both Th and B cells, was required at least to initiate the differentiation of CH12 cells. It was further shown that the interaction between CH12 and Th cells was MHC restricted. These requirements for MHC restricted Th cells and specific antigen had previously been described as typical of "resting" as opposed to "activated" or "excited" normal B cells. Detailed analysis of the MHC restriction of T helper cells for CH12 revealed that H-2 identity at I-E was both required and sufficient. Matching at I-A was not effective, even though CH12 expresses I-A which will activate alloreactive T cells to produce IL-2. The I-A restricted helper cells could provide help to normal I-A matched splenic B cells. These data suggested the possibility that binding of the I-E molecule by the Th cell might transmit a differentiative signal to the B cell, but that similar binding of the I-A molecule might not. We will return to this point.

a. *Role of Surface Immunoglobulin in B Cell Activation*

An important question in immunology has been answered using the CH12 lymphoma. The role of the membrane immunoglobulin on B lymphocytes has been debated for a number of years. Two theories were advanced: (1) the mIg molecules functioned only to focus T cell help on the B cell membrane,[157] or (2) the binding of antigen to mIg fulfills a signaling function.[159] To resolve this question requires that Th cell and B cell specificies for antigen be dissociated. This approach has been accomplished using CH12 and Th cells with specificity for either Ia^k or KLH (keyhole limpet hemocyanin).[159] CH12 cells were cultured with alloreactive Th cells with or without antigen. CH12 responded by differentiating to antibody secretion when T cells and specific antigen (SRBC) were present. Very high concentrations of Th cells alone were able to activate CH12, but only 1% as many T cells were required in cultures which contained SRBC. Culture of CH12 with Th cells specific for H-2^b encoded determinants (not on CH12) and a source of H-2^b antigen which was shown to stimulate the Th cells did not activate CH12 even in the simultaneous presence of SRBC. Th cells generated against KLH could also trigger CH12 to differentiate, but only if both KLH and SRBC were present in the culture. Separate experiments demonstrated that CH12 was able to present KLH antigen to the Th cells in the context of either I-A or I-E. The KLH Th cell-CH12 interaction leading to differentiation of CH12 was also restricted to the I-E subregion of the MHC. Taken together, the data indicate that activation of Th cells alone is not sufficient for their role in the activation of CH12 and that a direct interaction between Th cells and H-2-encoded determinants on CH12 cells is required for B cell activation to occur. The results show that linked interactions between the Th cells and B cells with specificity for epitopes on the same antigen molecule were not required for B cell activation to Ig secretion. This may, however, be the most efficient mechanism in vivo. These data also suggest a signaling function for the class II (Ia) molecules on the B cell membrane.

The latter point has been investigated.[160] Alloreactive Th cells, specific for either I-A^k or I-E^k, were used to activate CH12 cells in the presence of SRBC. Only the anti-I-E^k Th provided an activation signal. As with the system employing KLH-specific Th, both antigens which reacted with the mIg of CH12 and Th reactive with the I-E^k molecule were needed for optimal activation, but the requirement for antigen could be overwhelmed by very large numbers of T cells. These data confirmed that binding of antigen to mIg results in generation of a transmembrane differentiative signal. In addition, it was once again found that Th reactive with I-A^k, although themselves stimulated by CH12 to produce IL-2, and although demonstrably capable of inducing differentiation of normal B cells, could not cause CH12 to differentiate. This confirmed that a difference exists in the ability of I-A and I-E molecules of CH12 to transduce a differentiative signal across the membrane. Furthermore, since there was not any apparent difference between the behavior of Th cells activated by I-A^k and I-E^k, but only those reactive with I-E^k-stimulated CH12, it raised the possibility that the role of Th cells in generating an effective differentiative signal might not involve diffusible products of the T cells, but might require only that they bind to an active receptor (in this case I-E) on the B cell surface. Attempts were then made to activate CH12 with SRBC and monoclonal antibodies against either I-A^k or I-E^k. Anti-I-E^k and SRBC were able to induce CH12 to differentiate and secrete antibody, but anti-I-A^k and SRBC did not. Furthermore, antiidiotype antibody was successful in replacing the requirement for antigen.[160]

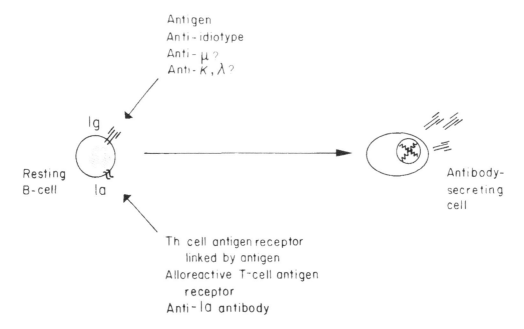

FIGURE 2.   Minimal model for B cell differentiation. This model is derived from observations of the
differentiation of the B cell lymphoma, CH12. Although the validity of the model for B cells, in general,
has not yet been tested, the model is consistent with most of the observations of antigen-driven differentiation of mature resting B cells. The model seeks only to describe the induced differentiation of B cells and
does not address the mechanisms of induced proliferation. A role for T cells and/or T cell factors in the
regulation, maintenance, and expansion of the response is not excluded by the model.

b. Minimal Model for B Cell Differentiation

These observations suggest the minimal model for B cell differentiation shown in
Figure 2. In this model the signals required to induce a B cell to differentiate to an
antibody secreting cell are binding of the surface immunoglobulin and signal-transduc-
ing Ia molecules. Occupation of these sites may be achieved in several ways, although
the cooperation of T cells linked to the B cell via antigen is probably the normal physiological mechanism. In this model, antigen binds specifically to the mIg of the B cell,
generating one part of the dual differentiative signal. This binding of antigen to the
cell surface then allows the formation of loose complexes between Ia molecules and
antigen, generating the requisite structures for recognition by the Th cell receptors.
The consequential binding of Ia generates the second part of the dual signal and leads
to differentiation of the B cell, without requiring participation of any metabolic product of the Th cell. Note that this model satisfies the basic requirements of antigen-
specific, MHC-restricted, T cell-dependent activation of B cells. The model explains
the requirement for linked recognition of antigen under normal conditions, but shows
how this may be by-passed in the presence of an allogeneic effect. The model implies
that binding of specific antigen is involved twice, generating both components of the
dual differentiative signal. This is compatible with ideas of the selective role of antigen
in stimulating antibody production, and in further selection of somatic mutants of Ig
which have an increased affinity for antigen. Note also that the model only seeks to
describe the induced differentiation of B cells and does not address the mechanisms of
induced proliferation. We have not felt that proliferative signals could be investigated
satisfactorily by the use of a neoplastic cell which, by its very nature, displays a rapid
constitutive rate of cell division. Certainly, these minimal differentiative signals are not
enough to achieve a functional serum antibody response. Proliferation and expansion
of antigen-specific B cell clones are required to generate enough antibody-secreting

cells, and it is likely that this is supported, maintained, and regulated by T cells or by the various T cell factors which have been described.

C. Differentiation of Plasmacytomas

Plasmacytomas, in addition to their usefulness in providing homogeneous sources of immunoglobulins, have provided an interesting model system for studying immunoregulation. Some fascinating work has been done by Lynch and co-workers[161,162] on the immunoregulation of the MOPC-315 myeloma. The MOPC-315 myeloma arose in a BALB/c mouse following injection of mineral oil.[35] This plasmacytoma produces an IgAλ_2 anti-TNP (trinitrophenyl) antibody and has been shown to differentiate in vivo.[161] During the early part of tumor growth there appears to be a high number of "stem-cell", nonsecretory lymphoid cells. About 10% of the cells at this stage secrete antibody. During growth of the tumor this early phenotype changes, so that in the latter stages of growth about 50% of the cells secrete antibody.

Early studies showed that immunization of BALB/c mice with the purified MOPC-315 myeloma protein, M315, would elicit production of antiidiotype antibodies.[163] These antibodies were specific for the MOPC-315 protein and would not cross react with the other myeloma proteins tested, even those which also had TNP specificity, such as MOPC-460. BALB/c mice immunized with the MOPC-315 protein also showed a resistance to challenge with a lethal dose of MOPC-315 cells.[164]

Later studies also showed that MOPC-315 cells responded to the presence of the TNP carrier antigen in the animal bearing the tumor.[165] By immunizing mice with different doses of carrier antigen, either carrier-specific help or suppression could be induced. Growth of MOPC-315 was enhanced in mice which had helper activity and inhibited in mice with suppressor activity. Furthermore, the regulatory effects in vivo did not require direct myeloma-regulatory cell interaction, since the myeloma cells were enclosed in diffusion chambers. Not only did the presence of regulatory cells modify the growth of MOPC-315, but differentiation was also modified. MOPC-315 cells grown in mice immunized with a suppressive dose of antigen had a greatly reduced number of anti-TNP PFC. TNP-rosette-forming cells were also decreased in mice with suppressive activity, indicating few cells expressing membrane M315 antibody.[165] The regulatory effects were later shown to be mediated by carrier-specific T cells and the helper and suppressor activities were contained in different T cell subpopulations.[166,167] Very recent evidence suggests that two types of helper cells were responsible for the positive immunoregulation of MOPC-315. Rohrer et al.[168] have reported that one type of T helper cell induces myeloma growth and clonal expansion and another is responsible for secretory differentiation. The first type, T_H1, is the classical MHC-restricted, carrier-specific Th cell. The latter type, T_H2, is the variously called idiotype, allotype, or isotype-specific Th. Using the MOPC-315 system as a model, the authors report that the T_H2 "sees" both Ig determinants as well as nominal antigen. Further, they find that T_H2 cells bear Qa-1, while the T_H1 cells do not.

The studies showing that immunization of BALB/c mice with M315 immunoglobulin could suppress tumor growth produced the finding[164] that some of the mice grew tumors, but in most of these mice there was no detectable serum M315. This suppression of M315 secretion was also observed if the MOPC-315 cells were enclosed in a diffusion chamber[167] and was reversed by removing the MOPC-315 cells to unimmunized recipients. The effect could be adoptively transferred to nonimmune mice with splenic T cells of M315 immune mice.[169] In all of these studies the M315 idiotype specificity of the effect was demonstrated in reciprocal experiments with MOPC-460 cells and/or M460 proteins.

In mice bearing MOPC-315, a large number of circulating small lymphocytes was found to bear the M315 protein on their membrane.[170] By growing MOPC-315 in appropriate F_1 mice, it was demonstrated that these cells are of F_1 (host) origin and not MOPC-315 cells. Further study showed these cells to be host T cells which appeared to have passively acquired the M315 produced by the tumor. That this was so was shown by Hoover and Lynch.[171] Using a number of IgA-secreting myelomas they demonstrated elevated numbers of T cells with mIgA. This mIgA could be competitively displaced with another purified IgA myeloma protein, but not with $F(ab')_2$ fragments. Release of the T cell membrane IgA exposed a surface membrane receptor for IgA, but not other Ig isotypes, and other Ig isotypes could not inhibit the IgA binding. Nonsecreting myeloma variants did not induce T cells with IgA binding ability. These T cells were shown to be Lyt-1⁻, 2⁺ and evidence was obtained that the induction of the Tα cells was dependent on the high level of circulating Ig and required DNA and protein synthesis.[172,173]

In 1982, Milburn and Lynch[174] described an in vitro system to study the T cell-mediated suppression of M315 secretion. They found that the inhibition of M315 secretion was a result of inhibition of M315 synthesis and that cell growth and viability were not affected. The effector cell was an Lyt-1⁻, 2⁺ T cell which recognized and specifically bound the M315 protein. Suppression was mediated via a 50KD soluble product and direct cell contact was not required. The T cell membrane receptor recognizes an idiotype located in the V_H region of M315.[175] Examination of immunoglobulin mRNA levels in suppressed and nonsuppressed MOPC-315 cells showed that λ_{II} light-chain transcription was inhibited.[176] In contrast, α heavy-chain mRNA levels were not inhibited and could be translated in a cell-free translation system.[177] In addition, another interesting observation was made. In suppressed MOPC-315 cells, not only was the normally expressed λ_{II} transcription inhibited, but the transcription of the unexpressed, aberrantly rearranged λ_I gene was also inhibited.[176] Therefore, it appears that although the idiotype regulation of MOPC-315 requires V_H encoded idiotype recognition by a T cell product, the actual mechanism operates via an inhibition of light-chain transcription.

VII. CONCLUSION

Malignant B cells have proven very useful in helping to understand the process of normal B lymphocyte activation. It can be argued that, since these cells are abnormal with respect to their growth, data derived from their study are unlikely to have relevance for understanding the biology of normal B cells. However, this argument is losing force in the face of many studies which demonstrate that these neoplastic B cells reflect, with remarkable fidelity, the behavior of normal B cells. It might be remembered that the initial studies using plasmacytoma-derived immunoglobulins to study antibody structure and function were criticized, because these immunoglobulins were abnormal "paraproteins". Likewise, similar criticisms have been raised about the initial study of immunoglobulin gene structure and regulation. With few exceptions, the subsequent analysis of normal B cells, using as a guide the information derived from their malignant counterparts, has demonstrated the accuracy of the data derived from neoplastic cells. These cells should continue to prove useful in the further elucidation of mechanisms controlling B cell differentiation.

REFERENCES

1. Early, R. and Hood, L., Mouse immunoglobulin genes, in *Genetic Engineering*, Vol. 3, Setlow, J. K. and Hollander, A., Eds., Plenum Press, New York, 1981, 157.

2. Saito, H., Kranz, D. M., Takagaki, Y., Hayday, A. C., Eisen, H. N., and Tonegawa, S., Complete primary structure of a heterodimeric T-cell receptor deduced from cDNA sequences, *Nature, (London)*, 309, 757, 1984.

3. Cantor, H. and Weissman, I., Development and function of subpopulations of thymocytes and T lymphocytes, *Prog. Allergy*, 20, 1, 1976.

4. Stutman, O., Intrathymic and extrathymic T cell maturation, *Transplant. Rev.*, 42, 138, 1978.

5. Ruddle, N. H., T cell tumors, clones, and hybrids, *Prog. Allergy*, 29, 222, 1981.

6. Till, J. E. and McCulloch, E. A., A direct measure of the radiation sensitivity of normal mouse bone marrow cells, *Radiat. Res.*, 14, 213, 1961.

7. Edwards, S. E., Miller, R. S., and Phillips, R. A., Differentiation of rosette-forming cells from myeloid stem cells, *J. Immunol.*, 105, 719, 1970.

8. Owen, J. J. T. and Jenkinson, E. J., Embryology of the lymphoid system, *Prog. Allergy*, 29, 1, 1981.

9. Osmond, D. S. and Nossal, G. J. V., Differentiation of lymphocytes in mouse bone marrow. II. Kinetics of maturation and renewal of antiglobulin-binding cells studied by double labelling, *Cell. Immunol.*, 13, 132, 1974.

10. Joho, R., Nottenburg, C., Coffman, R. L., and Weissman, I. L., Immunoglobulin gene rearrangement and expression during lymphocyte development, in *Current Topics in Developmental Biology*, Vol. 18, Moscona, A. A. and Mowray, A., Eds., Academic Press, New York, 1983, 15.

11. Möller, G., *Immunol. Rev.*, 59, 1981.

12. Coffman, R. L., Surface antigen expression and immunoglobulin gene rearrangement during mouse pre-B cell development, *Immunol. Rev.*, 69, 5, 1982.

13. Coffman, R. L. and Weissman, I. L., Immunoglobulin gene rearrangement during pre-B cell differentiation, *J. Mol. Cell. Immunol.*, 1, 31, 1983.

14. Hammerling, N., Chin, A. F., and Abbott, J., Ontogeny of murine B lymphocytes. Sequence of B cell differentiation from surface-immunoglobulin negative precursors to plasma cells, *Proc. Natl. Acad. Sci. U.S.A.*, 73, 2008, 1976.

15. Shen, F. W., Yakura, H., and Tung, J. S., Some compartments of B cell differentiation, *Immunol. Rev.*, 69, 69, 1983.

16. Subbarao, B. and Mosier, D. E., Lyb antigens and their role in B lymphocyte activation, *Immunol. Rev.*, 69, 81, 1983.

17. Kemp, J. D., Rohrer, J. W., and Huber, B. T., Lyb3: a B cell surface antigen associated with triggering secretory differentiation, *Immunol. Rev.*, 69, 127, 1983.

18. McKearn, J. P., Paslay, J. W., Slack, J., Baum, C., and Davie, J. M., B cell subsets and differential responses to mitogens, *Immunol. Rev.*, 64, 5, 1982.

19. Hardy, R. R., Hayakawa, K., Parks, D. R., Hergenberg, L. A., and Herzenberg, L. A., Murine B cell differentiation lineages, *J. Exp. Med.*, 159, 1169, 1984.

20. Kung, J. T. and Paul, W. E., B-lymphocyte subpopulations, *Immunol. Today*, 4, 37, 1983.

21. McKearn, J. P., Baum, C., and Davie, J. M., Cell surface antigens expressed by subsets of pre-B cells and B cells, *J. Immunol.*, 132, 332, 1984.

22. Kronenberg, M., Davis, M. M., Early, P. W., Hood, L. E., and Watson, J. D., Helper and killer T cells do not express B cell immunoglobulin joining and constant region segments, *J. Exp. Med.*, 152, 1745, 1980.

23. Wetzel, G. D. and Kettman, J. R., Activation of murine B lymphocytes. III. Stimulation of B lymphocyte clonal growth with lipopolysaccharide and dextran sulphate, *J. Immunol.*, 126, 723, 1981.

24. Melchers, F., Anderson, J., and Phillips, R. A., Ontogeny of murine B lymphocytes: development of Ig synthesis and of reactivities to mitogens and anti-Ig antibodies, *Cold Spring Harbor Symp. Quant. Biol.*, 41, 147, 1976.

25. Cooper, M. D., Kearney, J. F., Gathings, W. E., and Lawton, A. R., Effects of anti-Ig antibodies on the development and differentiation of B cells, *Immunol. Rev.*, 52, 19, 1980.

26. Parker, D. C., Induction and suppression of polyclonal antibody responses by anti-Ig reagents and antigen-nonspecific helper factors, *Immunol. Rev.*, 52, 115, 1980.

27. Sieckman, D. G., The use of anti-immunoglobulins to induce a signal for cell division in B lymphocytes via their membrane IgM and IgD, *Immunol. Rev.*, 52, 181, 1980.

28. Vitetta, E., Pure, E., Isakson, P., Buck, L., and Uhr, J., The activation of murine B cells: the role of surface immunoglobulin, *Immunol. Rev.*, 52, 211, 1980.

29. Galsebrook, A. L., Sarmiento, M., Loken, M. R., Dialynas, D. P., Quintans, J., Eisenberg, L., Lutz, C. T., Wilde, D., and Fitch, F. W., Murine T lymphocyte clones with distinct immunological functions, *Immunol. Rev.*, 54, 225, 1981.

30. Sredni, B. and Schwartz, R. H., Antigen-specific, proliferating T lymphocyte clones. Methodology, specificity, MHC restriction and alloreactivity, *Immunol. Rev.*, 54, 187, 1981.

31. Howard, M., Kessler, S., Chused, T., and Paul, W. E., Long term culture of normal mouse B lymphomas, *Proc. Natl. Acad. Sci. U.S.A.*, 78, 5788, 1981.

32. Aldo-Benson, M. and Scheider, L., Long-term growth of lines of murine dinitrophenyl-specific B lymphocytes in vitro, *J. Exp. Med.*, 157, 342, 1983.

33. Lanier, L. L., Warner, N. L., Ledbetter, J. A., and Herzenberg, L. A., Quantitative immunofluorescent analysis of surface phenotypes of murine B cell lymphomas and plasmacytomas with monoclonal antibodies, *J. Immunol.*, 127, 1691, 1981.

34. Makela, O. and Cross, A. M., The diversity and specialization of immunocytes, *Prog. Allergy*, 14, 145, 1970.

35. Potter, M., Immunoglobulin-producing tumors and myeloma proteins of mice, *Physiol. Rev.*, 52, 631, 1972.

36. Warner, N. L., Autoimmunity and the pathogenesis of plasma cell tumor induction in NZB inbred and hybrid mice, *Immunogenetics*, 2, 1, 1975.

37. Potter, M., Pumphrey, J. G., and Bailey, D. W., Genetics of susceptibility to plasmacytoma induction. I. Balb/cAnN(c), C57BL/6N(B6), C57BL/Ka(BK), (C×B6)F$_1$, (C×BK)F$_1$, and C×B recombinant-inbred strains, *J. Natl. Cancer Inst.*, 54, 1413, 1975.

38. Morse, H. C., Riblet, R., Asofsky, R., and Weigert, M., Plasmacytomas of the NZB mouse, *J. Immunol.*, 121, 1969, 1978.

39. Loh, E., Hood, J. M., Riblet, R., Weigert, M., and Hood, L., Comparisons of myeloma proteins from NZB and Balb/c mice: structural and functional differences of heavy chains, *J. Immunol.*, 122, 44, 1979.

40. Loh, E., Black, B., Riblet, R., Wiegert, M., Hood, J. M., and Hood, L., Myeloma proteins from NZB and Balb/c mice: structural and functional differences, *Proc. Natl. Acad. Sci. U.S.A.*, 76, 1395, 1979.

41. Scharff, M. D. and Laskov, R., Synthesis and assembly of immunoglobulin polypeptide chains, *Prog. Allergy*, 14, 37, 1970.

42. Milstein, C. and Pink, J. R. L., Structure and evolution of immunoglobulins, *Prog. Biophys. Mol. Biol.*, 21, 209, 1970.

43. Moller, G., *Immunol. Rev.*, 67, 1982.

44. Mathe, G., Seligmann, M., and Tubiana, M., Eds., Lymphoid neoplasms I. Classification, categorization and natural history, in *Recent Results in Cancer Research*, Vol. 64, Springer-Verlag, New York, 1978.

45. Abelson, H. T. and Robstein, L. S., Lymphoma sarcoma: virus-induced thymic-independent disease in mice, *Cancer Res.*, 30, 2213, 1970.

46. Raschke, W. C., Ralph, P., Watson, J., Sklar, M., and Coon, H., Oncogenic transformation of murine lymphoid cells by in vitro infection with Abelson leukemia virus, *J. Natl. Cancer Inst.*, 54, 1249, 1975.

47. Paige, C. J., Kincade, P. W., and Ralph, P., Murine B cell leukemia line with inducible surface immunoglobulin expression, *J. Immunol.*, 121, 641, 1978.

48. Dessner, D. S. and Loken, M. R., DNL 1.9: a monoclonal antibody which specifically detects all murine B lineage cells, *Eur. J. Immunol.*, 11, 282, 1981.

49. Tada, N., Kimura, S., Liu, Y., Taylor, B. A., and Hammerling, U., Ly-m19: the Lyb-2 region of mouse chromosome 4 controls a new surface alloantigen, *Immunogenetics*, 13, 529, 1981.

50. Kimura, S., Tada, N., Nakayama, E., Liu, Y., and Hammerling, U., A new mouse cell-surface antigen (Ly-m20) controlled by a gene linked to Mls locus and defined by monoclonal antibodies, *Immunogenetics*, 14, 3, 1981.

51. Coffman, R. L. and Weissman, I. L., A monoclonal antibody that recognizes B cells and B cell precursors in mice, *J. Exp. Med.*, 153, 269, 1981.

52. Coffman, R. L. and Weissman, I. L., B220: a B cell-specific member of the T200 glycoprotein family, *Nature (London)*, 289, 681, 1981.

53. Sitia, R., Rubartelli, A., and Hammerling, U., Expression of 2 immunoglobulin isotypes, IgM and IgA, with identical idiotype in the B cell lymphoma I.29, *J. Immunol.*, 127, 1388, 1981.

54. Sato, H. and Boyse, E., A new alloantigen expressed selectively on B cells: the Lyb-2 system, *Immunogenetics*, 3, 565, 1976.

55. McKenzie, I. F. C. and Zola, H., Monoclonal antibodies to B cells, *Immunol. Today*, 4, 10, 1983.

56. Subbarao, B. and Mosier, D. E., Lyb antigens and their role in B lymphocyte activation, *Immunol. Rev.*, 69, 81, 1982.

57. Yutoku, M., Grossberg, A. L., and Pressman, D., A cell surface antigenic determinant present on mouse plasmacytes and only about half of mouse thymocytes, *J. Immunol.*, 112, 1774, 1974.

58. Eckhardt, L. A. and Herzenberg, L. A., Monoclonal antibodies to ThB detect close linkage of Ly-6 and a gene regulating ThB expression, *Immunogenetics*, 11, 275, 1980.

59. Takahashi, T., Old, L. J., and Boyse, E. A., Surface alloantigens of plasma cells, *J. Exp. Med.*, 131, 1325, 1970.

60. Tada, N., Kimura, S., Hoffman, M., and Hammerling, U., A new surface antigen (PC.2) expressed exclusively on plasma cells, *Immunogenetics*, 11, 351, 1980.

61. Warner, N. L., Daley, M. J., Richey, J., and Spellman, C., Flow cytometry analysis of murine B cell lymphoma differentiation, *Immunol. Rev.*, 48, 197, 1979.

62. Lanier, L. L., Walker, E. B., Richie, E. R., and Warner, N. L., Murine hematopoietic cell tumors: models for analysis of cellular differentiation, in *B and T Cell Tumors*, Vitetta, E. S., Ed., Raven Press, New York, 1982, 223.

63. Warner, N. L., Cheney, R. K., Lanier, L. L., Daley, M., and Walker, E., Differentiation heterogeneity in murine hematopoietic tumors, in *Maturation Factors and Cancer*, Moore, M. A. S., Ed., Raven Press, New York, 1982, 223.

64. Godal, T., Lindmo, R., Ruud, E., Heikkila, R., Henriksen, A., Steen, H. B., and Marton, P. F., Immunologic subsets in human B-cell lymphomas in relation to normal B-cell development, *Haematol. Blood Transfus.*, 26, 314, 1981.

65. Godal, T., Lindmo, R., Marton, P. F., Landaas, T. O., Langholm, R., Hoie, J., and Abrahamsen, A. F., Immunologic subsets in human B-cell activation, *Scand. J. Immunol.*, 14, 481, 1981.

66. Godal, T. and Funderud, S., Human B-cell neoplasms in relation to normal B cell differentiation and maturation processes, *Adv. Cancer Res.*, 36, 211, 1982.

67. Funderud, S., Lindmo, T., Ruud, E., Marton, P. F., Langholm, R., Elgio, R. F., Vaage, S., Liu, S., and Godal, T., Delineation of subsets in human B-cell lymphomas by a set of monoclonal antibodies raised against B lymphoma cells, *Scand. J. Immunol.*, 17, 161, 1983.

68. Melsom, H., Funderud, S., Liu, S. O., and Godal, T., The great majority of childhood lymphoblastic leukemias are identical by monoclonal antibodies as neoplasms of the B-cell progenitor compartment, *Scand. J. Haematol.*, 33, 27, 1984.

69. Ledbetter, J. A. and Herzenberg, L. A., Xenogeneic monoclonal antibodies to mouse lymphoid antigens, *Immunol. Rev.*, 47, 64, 1979.

70. Lanier, L. L. and Warner, N. L., Cell cycle related heterogeneity of Ia antigen expression on a murine B lymphoma cell line: analysis of flow cytometry, *J. Immunol.*, 126, 626, 1981.

71. Warner, N. L., Neoplasms of immunoglobulin producing cells in mice, *Rec. Results Cancer Res.*, 64, 316, 1978.

72. Cantor, H. and Boyse, E. A., Lymphocytes as models for the study of mammalian cellular differentiation, *Immunol. Rev.*, 33, 105, 1971.

73. Mathieson, B. T., Sharrow, S. O., Campbell, P. S., and Asofsky, R., An Lyt differentiated subpopulation of thymocytes detected by flow cytometry, *Nature (London)*, 277, 478, 1979.

74. Ledbetter, J. A., Rouse, R. V., Micklem, H. S., and Herzenberg, L. A., T cell subsets defined by expression of Lyt-1,2,3 and Thy-1 antigens: two parameter immunofluorescence and cytotoxicity analysis with monoclonal antibodies modified current views, *J. Exp. Med.*, 152, 280, 1980.

75. Lanier, L. L., Warner, N. L., Arnold, L. W., Raybourne, R. B., and Haughton, G., B cell lymphomas in New Zealand and B10.H-2ᵃH-4ᵇp/Wts mice: antigenic analysis by flow cytometry, in *B Lymphocytes in the Immune Response: Functional, Developmental, and Interactive Properties*, Klinman, N., Mosier, D. E., Scher, I., and Vitetta, E. S., Eds., Elsevier/North-Holland, New York, 1981, 459.

76. Lanier, L. L., Warner, N. L., Ledbetter, J. A., and Herzenberg, L. A., Expression of Lyt-1 antigen on certain murine B cell lymphomas, *J. Exp. Med.*, 998, 1981.

77. Lanier, L. L., Arnold, L. W., Raybourne, R. B., Russell, S., Lynes, M. A., Warner, N. L., and Haughton, G., Transplantable B-cell lymphomas in B10.H-2ᵃH-4ᵇp/Wts mice, *Immunogenetics*, 16, 367, 1982.

78. Ledbetter, J. A., Evans, R. L., Lipinski, M., Cunningham-Rundles, C., Good, R. A., and Herzenberg, L. A., Evolutionary conservation of surface molecules that distinguish T lymphocyte helper/inducer and cytotoxic/suppressor subpopulations in mouse and man, *J. Exp. Med.*, 153, 310, 1981.

79. Wang, C. Y., Good, R. A., Ammirato, P., Dymbart, G., and Evans, R. L., Identification of a p69/70 complex expressed on human T cells sharing determinants with B-type chronic lymphatic leukemia cells, *J. Exp. Med.*, 151, 1539, 1980.

80. Burns, B. F., Warnke, R. A., Doggett, R. S., and Rouse, R. V., Expression of a T-cell antigen (Leu-1) by B-cell lymphomas, *Am. J. Pathol.*, 113, 165, 1983.

81. Hardy, R. R., Hayakawa, K., Haaijman, J., and Herzenberg, L. A., B cell subpopulations identifiable by two-color fluorescence analysis using a dual-laser FACS, *Ann. N.Y. Acad. Sci.*, 399, 112, 1982.

82. Manohar, V., Brown, E., Leiserson, W. M., and Chused, T. M., Expression of Lyt-1 by a subset of B lymphocytes, *J. Immunol.*, 129, 532, 1982.

83. Hayakawa, K., Hardy, R. R., Parks, D. R., and Herzenberg, L. A., The "Ly-1 B" cell subpopulation in normal, immunodefective and autoimmune mice, *J. Exp. Med.*, 151, 202, 1983.

84. Okumura, K., Hayakawa, K., and Tada, T., Cell-to-cell interaction controlled by immunoglobulin genes. Role of Thy-1⁻, Lyt-1⁺ (B′) cell in allotype-restricted antibody production, *J. Exp. Med.*, 156, 443, 1982.

85. Hayakawa, K., Hardy, R. R., Honda, M., Herzenberg, L. A., Steinberg, A., and Herzenberg, L. A., Ly-1 B cells: functionally distinct lymphocytes that secrete IgM autoantibodies, *Proc. Natl. Acad. Sci. U.S.A.*, 81, 2494, 1984.

86. Lanier, L. L., Richie, E. R., Howell, A. L., and Allison, J. P., Expression of Ly-1 and Ly-2 on a spontaneous AKR B-cell lymphoma, *Immunogenetics*, 17, 655, 1983.

87. Greenberg, R. S., Mathieson, B. J., Campbell, P. S., and Zatz, M. M., Multiple occurrence of spontaneous AKR/J lymphomas with T and B cell characteristics, *J. Immunol.*, 118, 1181, 1977.

88. Zatz, M. M., Mathieson, B. J., Kanellopoulos-Langevin, C., and Sharrow, S. O., Separation and characterization of two component tumor lines within the AKR lymphoma, AKTB-1, by fluorescence-activated cell sorting and flow cytometry analysis. I. The coexistence of sIg⁺ and sIg⁻ sublines, *J. Immunol.*, 126, 608, 1981.

89. Pennell, C. A., Arnold, L. W., Lutz, P. M., LoCascio, N. J., Willoughby, P. B. and Haughton, G., Cross-reactive idiotypes and common antigen binding specificities expressed by a series of murine B cell lymphomas: etiological implications, *Proc. Natl. Acad. Sci. U.S.A.*, 82, 1985.

90. Fu, S. M., Chiorazzi, N., and Kunkel, H. G., Differentiation capacity and other properties of the leukemic cells of chronic lymphocytic leukemia, *Immunol. Rev.*, 48, 23, 1979.

91. Kishimoto, T., Activation of human B-lymphoblastoid cell lines to IgG-producing cells by allogeneic T cells, in *B Lymphocytes in the Immune Response*, Cooper, M., Mosier, D. E., Scher, I., and Vitetta, E. S., Eds., Elsevier/North-Holland, New York, 1979, 285.

92. Robert, K.-H., Induction of monoclonal antibody synthesis in malignant human B cells by polyclonal B cell activators, *Immunol. Rev.*, 48, 123, 1979.

93. Ralph, P., Functional subsets of murine and human B lymphocyte cell lines, *Immunol. Rev.*, 48, 107, 1979.

94. Baltimore, D., Rosenberg, N., and Witte, O. N., Transformation of immature lymphoid cells by Abelson murine leukemia virus, *Immunol. Rev.*, 48, 3, 1979.

95. Rosenberg, N., Siden, E., and Baltimore, D., Synthesis of μ chains by Abelson virus-transformed cells and induction of light chain synthesis with lipopolysaccharide, in *B Lymphocytes in the Immune Response*, Cooper, M., Mosier, D. E., Scher, I., and Vitetta, E. S., Eds., Elsevier/North-Holland, New York, 1979, 379.

96. Siden, E. J., Baltimore, D., Clark, D., and Rosenberg, N., Immunoglobulin synthesis by lymphoid cells transformed in vitro by Abelson murine leukemia virus, *Cell*, 16, 389, 1979.

97. Riley, S. C., Brock, E. J., and Kuehl, W. M., Induction of light chain expression in a pre-B cell line by fusion to myeloma cells, *Nature (London)*, 289, 804, 1981.

98. Baines, P., Dexter, T. M., and Schofield, R., Characterization of malignant cell populations in MNU-induced leukemia of mice, *Leukemia Res.*, 3, 23, 1979.

99. Maki, R., Kearney, J., Paige, C., and Tonegawa, S., Immunoglobulin gene rearrangement in immature B cells, *Science*, 209, 1366, 1980.

100. Ralph, P., Nakoinz, I., and Roschke, W. C., Lymphosarcoma cell growth is selectively inhibited by B lymphocyte mitogens: LPS, dextran sulfate and PPD, *Biochem. Biophys. Res. Commun.*, 61, 1268, 1974.

101. Boss, M., Greaves, M., and Teich, N., Abelson virus-transformed hematopoietic cell lines with pre-B cell characteristics, *Nature (London)*, 278, 551, 1979.

102. Teale, J. M., The relationship between pre-B cells and primary B cells, *Immunol. Today*, 3, 62, 1982.

103. Paige, C. J., Kincade, P. W., and Ralph, P., Independent control of immunoglobulin heavy and light chain expression in a murine pre-B cell line, *Nature (London)*, 292, 631, 1981.

104. Parslow, T. G. and Granner, D. K., Chromatin changes accompanying immunoglobulin kappa gene activation: a potential control region within the gene, *Nature (London)*, 299, 449, 1982.

105. Giri, J. G., Kincade, P. W., and Mizel, S. B., Interleukin 1-mediated induction of x-light chain synthesis and surface immunoglobulin expression on pre-B cells, *J. Immunol.*, 132, 223, 1984.

106. Paige, C. J., Schreirer, M. H., and Sidman, C. L., Mediators from cloned T helper cell lines affect immunoglobulin expression by B cells, *Proc. Natl. Acad. Sci. U.S.A.*, 79, 4756, 1982.

107. Fu, S. M., Winchester, R. J., and Kunkel, H. G., Occurrence of surface IgM, IgD and free light chains on human lymphocytes, *J. Exp. Med.*, 139, 451, 1974.

108. Kubo, R. T., Grey, H. M., and Pirofsky, B., IgD: a major immunoglobulin on the surface of lymphocytes from patients with chronic lymphatic leukemia, *J. Immunol.*, 112, 1952, 1974.

109. Seligman, M., Preud'Homme, J. -L., and Brouet, J. -C., B and T cell markers in human proliferative blood diseases and primary immunodeficiencies, with special reference to membrane bound immunoglobulins, *Transplant. Rev.*, 16, 85, 1973.

110. Fu, S. M., Winchester, R. J., and Kunkel, H. G., Similar idiotypic specificity for the membrane IgD and IgM of human B lymphocytes, *J. Immunol.*, 114, 250, 1975.

111. Salsano, F., Froland, S. S., Natvig, J. B., and Michaelson, T. E., Same idiotype of B lymphocyte membrane IgD and IgM. Formal evidence for monoclonality of chronic lymphocytic leukemia cells, *Scand. J. Immunol.*, 3, 841, 1974.

112. Johnstone, A. P., Chronic lymphocytic leukemia and its relationship to normal B lymphopoiesis, *Immunol. Today*, 3, 343, 1982.

113. Fu, S. M., Winchester, R. J., Feizi, T., Walzer, P. D., and Kundel, H. G., Idiotypic specificity of surface immunoglobulin and the maturation of leukemic bone-marrow derived lymphocytes, *Proc. Natl. Acad. Sci. U.S.A.*, 71, 4487, 1974.

114. Hurley, J. N., Fu, S. M., Kundel, H. G., McKenna, G., and Scharff, M. D., Lymphoblastoid cell lines from patients with chronic lymphocytic leukemia: identification of tumor origin by idiotypic analysis, *Proc. Natl. Acad. Sci. U.S.A.*, 75, 5706, 1978.

115. Godal, T., On the complexity of the B cell system as assessed by studies on human B cell neoplasms, in *B and T Cell Tumors*, Vitetta, E. S., Ed., Academic Press, New York, 1982, 69.

116. Warner, N. L., Harris, A. W., and Gutman, G. A., Membrane immunoglobulin and Fc receptors of murine T and B cell lymphomas, in *Membrane Receptors of Lymphocytes*, Seligman, M., Preud'Homme, J. -L., and Kourilsky, F. M., Eds., North-Holland, Amsterdam, 1975, 203.

117. Kim, K. J., Kanellopoulos-Langevin, C., Merwin, R. M., Sachs, D. H., and Asofsky, R., Establishment and characterization of Balb/c lymphoma lines with B cell properties, *J. Immunol.*, 122, 549, 1979.

118. Slavin, S. and Strober, S., Spontaneous murine B cell leukemia, *Nature (London)*, 272, 624, 1978.

119. Strober, S., Gronowicz, E. S., Knapp, M. R., Slavin, S., Vitetta, E. S., Warnke, R. A., Kotzin, B., and Schroder, J., Immunobiology of a spontaneous murine B cell leukemia (BCL₁), *Immunol. Rev.*, 48, 169, 1979.

120. Knapp, M. R., Jones, P. P., Black, S. J., Vitetta, E. S., Slavin, S., and Strober, S., Characterization of a spontaneous murine B cell leukemia (BCL₁). I. Cell surface expression of IgM, IgD, Ia, and FcR, *J. Immunol.*, 123, 992, 1979.

121. Haughton, G., Arnold, L. W., Lanier, L. L., Raybourne, R. B., and Warner, N. L., Induction of B cell lymphomas in B10.H-2ᵃH-4ᵇp/Wts mice, in *B Lymphocytes in the Immune Response: Functional, Developmental, and Interactive Properties*, Klinman, D., Mosier, D. E., Scher, I., and Vitetta, E. S., Eds., Elsevier/North-Holland, New York, 1981, 455.

122. Lanier, L. L., Arnold, L. W., Raybourne, R. B., Russell, S., Lynes, M. A., Warner, N. L., and Haughton, G., Transplantable B-cell lymphomas in B10.H-2ᵃH-4ᵇp/Wts mice, *Immunogenetics*, 16, 367, 1982.

123. Corley, R. B., Arnold, L. W., Lutz, P. M., Kuhara, T., White, D. A., Staniszewski, C., and Haughton, G., T cell lymphomas with helper T cell function, in *Intercellular Communication in Leucocyte Function*, Parker, J. W. and O'Brien, R. L., Eds., John Wiley & Sons, New York, 1983, 521.

124. Arnold, L. W., LoCascio, N. J., Lutz, P. A., Pennell, C. A., Klapper, D., and Haughton, G., Antigen-induced lymphomagenesis: identification of a murine B cell lymphoma with known antigen specificity, *J. Immunol.*, 131, 2064, 1983.

125. Arnold, L. W., Pennell, C. A., LoCascio, N. J., and Haughton, G., Cross-reactive idiotypes of murine B cell lymphomas: implications for etiology and immunotherapy, in *B and T Cell Tumors*, Vitetta, E. S., Ed., Academic Press, New York, 1982, 499.

126. Haskins, K., Kappler, J., and Marrack, P., The major histocompatibility complex-restricted antigen receptor on T cells, *Ann. Rev. Immunol.*, 2, 51, 1984.

127. Unanue, E. R., Antigen-presenting function of the macrophage, *Ann. Rev. Immunol.*, 2, 395, 1984.

128. Stingl, G., Katz, S. I., Green, I., and Shevach, E. M., The functional role of Langerhans cells, *J. Invest. Dermatol.*, 74, 315, 1980.

129. Steinman, R. M. and Nussenzweig, M. C., Dendritic cells: features and functions, *Immunol. Rev.*, 53, 127, 1980.

130. Chesnut, R. W. and Grey, H. M., Studies on the capacity of B cells to serve as antigen-presenting cells, *J. Immunol.*, 126, 1075, 1981.

131. Walker, E. B., Lanier, L. L., and Warner, N. L., Characterization and functional properties of tumor cell lines in accessor cell replacement assays, *J. Immunol.*, 128, 852, 1982.

132. Glimcher, L. H., Kim, K. -J., Green, I., and Paul, W. E., Ia antigen-bearing B cell tumor lines can present protein antigen and alloantigen in a major histocompatibility complex-restricted fashion to antigen-reactive T cells, *J. Exp. Med.*, 155, 445, 1982.

133. Glimcher, L. H., Hamano, T., Asofsky, R., Herber-Katz, E., Hedrick, S., Schwartz, R. H., and Paul, W. E., I region-restricted antigen presentation by B cell-B lymphoma hybridomas, *Nature (London)*, 298, 283, 1982.

134. Andersson, J., Coutinho, A., Lernhardt, W., and Melchers, F., Clonal growth and maturation to immunoglobulin secretion in vitro of every growth-inducible B lymphocyte, *Cell*, 10, 27, 1977.

135. Knapp, M. R., Gronowicz, E. S., Schroeder, J., and Strober, S., Characterization of the spontaneous murine B cell leukemia (BCL₁). II. Tumor cell proliferation and IgM secretion after stimulation by LPS, *J. Immunol.*, 123, 1000, 1979.

136. Gronowicz, E. S., Doos, C. A., Howard, F. D., Morrison, D. C., and Strober, S., An in vitro line of the B cell tumor BCL₁ can be activated by LPS to secrete IgM, *J. Immunol.*, 125, 975, 1980.

137. Boyd, A. W., Goding, J. W., and Schrader, J. W., The regulation of growth and differentiation of a murine B cell lymphoma. I. Lipopolysaccharide-induced differentiation, *J. Immunol.*, 126, 2461, 1981.

138. Lanier, L. L., Activation of murine B cell lymphomas. Influence of lipopolysaccharide, *J. Immunol.*, 129, 1130, 1982.

139. Ralph, P. and Nakoinz, I., Lipopolysaccharides inhibit lymphosarcoma cells of bone marrow origin, *Nature (London)*, 249, 49, 1974.

140. LoCascio, N. J., Arnold, L. W., Corley, R. B., and Haughton, G., Induced differentiation of a B cell lymphoma with known antigen specificity, *J. Mol. Cell Immunol.*, 1, 177, 1984.

141. LaFrenz, D., Koretz, S., Stratte, P. T., Ward, R. B., and Strober, S., LPS-induced differentiation of a murine B-cell leukemia (BCL₁): changes in surface and secreted IgM, *J. Immunol.*, 129, 1329, 1982.

142. Yuan, D. and Tucker, P. W., Effect of LPS stimulation on the transcription and translation of mRNA for cell surface IgM, *J. Exp. Med.*, 156, 962, 1982.

143. Yuan, D., Role of glycosylation in the cell surface expression and secretion of Ig molecules by BCL₁ cells, *Mol. Immunol.*, 19, 1149, 1982.

144. Isakson, P. C., Pure, E., Uhr, J. W., and Vitetta, E. S., Induction of proliferation and differentiation of neoplastic B cells by antiimmunoglobulin and T-cell factors, *Proc. Natl. Acad. Sci. U.S.A.*, 78, 2507, 1981.

145. Godal, T., Henriksen, A., Rund, E., and Michaelson, T., Monoclonal human B lymphoma cells respond by DNA synthesis to anti-immunoglobulin in the presence of the tumor promoter TPA, *Scand. J. Immunol.*, 12, 267, 1982.

146. Beiske, K., Ruud, E., Marton, P. F., and Godal, T., Induction of maturation of human B-cell lymphomas in vitro: morphologic changes in relation to immunoglobulin and DNA synthesis, *Am. J. Pathol.*, 115, 362, 1984.

147. Rudd, E., Steen, H. B., Beiske, K., and Godal, T., Different responses elicited in vitro by antibodies to IgM and IgD in cells from surface IgM and IgD positive human follicular lymphoma, *Scand. J. Immunol.*, 17, 155, 1983.

148. Godal, T., Rund, E., Heikkila, R., Funderud, S., Michaelsen, R., Jeffries, R., Ling, N. R., and Hildrum, K., Triggering of monoclonal human lymphoma B cells with antibodies to IgM heavy chains: differences of response obtained with monoclonal as compared to polyclonal antibodies, *Clin. Exp. Immunol.*, 54, 756, 1983.

149. Kishimoto, T., Yoshizaki, K., Okada, M., Miki, Y., Nakagawa, T., Yoshimura, N., Kishi, H., and Yamamura, Y., Activation of human monoclonal B cells with anti-Ig and T cell-derived helper factor(s) and biochemical analysis of transmembrane signaling in B cells, in *B and T Cell Tumors*, Vitetta, E. S., Ed., Academic Press, New York, 1982, 275.

150. Yoshizaki, K., Nakagawa, T., Kaieda, T., Muraguchi, A., Yamamura, Y., and Kishimoto, T., Induction of proliferation and Ig production in human B leukemic cells by anti-immunoglobulins and T cell factors, *J. Immunol.*, 128, 1296, 1982.

151. Pure, E., Isaksom, P. C., Paetkau, V., Caplain, B., Vitetta, E. S., and Krammer, P. H., Interleukin-2 does not induce murine B cells to secrete Ig, *J. Immunol.*, 129, 2420, 1982.

152. Muraguchi, A., Kishimoto, T., Miki, Y., Kuritani, T., Kaieda, T., Yoshizaki, K., and Yamamura, Y., T cell-replacing factor (TRF)-induced IgG secretion in a human B blastoid cell line and demonstration of acceptors for TRF, *J. Immunol.*, 127, 412, 1981.

153. Okada, M., Sakaguchi, N., Yoshimura, N., Hara, H., Shimizu, K., Yoshida, N., Yoshizaka, K., Kishimoto, S., Yamamura, Y., and Kishimoto, T., B cell growth factors and B cell differentiation factor from human T hybridomas. Two distinct kinds of B cell growth factor and their synergism in B cell proliferation, *J. Exp. Med.*, 157, 583, 1983.

154. Isakson, P. C., Krolick, K. A., and Vitetta, E. S., The effect of antiimmunoglobulin antibodies on the in vitro proliferation and differentiation of normal and neoplastic murine B cells, *J. Immunol.*, 125, 886, 1980.

155. Severinson, E., Torres, A., and Moller, G., Activation of BCL₁ cells by polyclonal activators and inhibition of growth and IgM secretion by anti-IgM,, *Scand. J. Immunol.*, 18, 153, 1983.

156. Arnold, L. W., LoCascio, N. J., Lutz, P. M., Pennell, C. A., Klapper, D., and Haughton, G., Antigen-induced lymphogenesis: identification of a murine B cell lymphoma with known antigen specificity, *J. Immunol.*, 131, 2064, 1983.

157. Coutinho, A. and Möller, G., Immune activation of B cells: evidence for "one nonspecific triggering signal" not delivered by the Ig receptor, *Scand. J. Immunol.*, 3, 133, 1974.

158. Bretscher, P. and Cohn, M., A theory of self-nonself discrimination, *Science,* 169, 1042, 1970.

159. LoCascio, N. J., Haughton, G., Arnold, L. W., and Corley, R. B., Role of cell surface immunoglobulin in B-lymphocyte activation, *Proc. Natl. Acad. Sci. U.S.A.,* 81, 2466, 1984.

160. LoCascio, N. J., Arnold, L. W., Haughton, G., and Corley, R. B., Anti-class II substitutes for Th cell driven differentiation of B cells, submitted.

161. Lynch, R. G., Rohrer, J. W., Adermatt, B., Gebel, H. M., Autry, J. R., and Hoover, R. G., Immunoregulation of murine myeloma cell growth and differentiation: a monoclonal model of B cell differentiation, *Immunol. Rev.,* 48, 45, 1979.

162. Lynch, R. G., Milburn, G. L., Urnovitz, H. B., Binion, S. B., Williams, K. R., and Mueller, A., The analyses of immunoregulatory mechanisms with myeloma, lymphoma, and hybridoma cells, *Progress in Immunology V,* Yamamura, Y. and Tada, T., Eds., Academic Press, New York, 1983, 719.

163. Sirisinha, S. and Eisen, H. N., Autoimmune antibodies to the ligand binding sites of myeloma proteins, *Proc. Natl. Acad. Sci. U.S.A.,* 68, 3130, 1971.

164. Lynch, R. G., Graff, R., Sirisinha, S., Simms, E. S., and Eisen, H. N., Myeloma proteins as tumor-specific transplantation antigens, *Proc. Natl. Acad. Sci. U.S.A.,* 69, 1540, 1972.

165. Rohrer, J. W. and Lynch, R. G., Specific immunologic regulation of differentiation of immunoglobulin expression in MOPC-315 cells during in vivo growth in diffusion chambers, *J. Immunol.,* 119, 2045, 1977.

166. Rohrer, J. W. and Lynch, R. G., Antigen-specific regulation of myeloma cell differentiation in vivo by carrier-specific T cell factors and macrophages, *J. Immunol.,* 120, 1066, 1978.

167. Rohrer, J. W., Adermatt, B. O., and Lynch, R. G., Idiotype-specific suppression of MOPC-315 IgA secretion in vivo: reversible blockade of secretory myeloma cells by soluble mediators, *J. Immunol.,* 121, 1799, 1978.

168. Rohrer, J. W., Gershon, R. K., Lynch, R. G., and Kemp, J. D., Enhancement of B lymphocyte secretory differentiation by a Ly1+, 2−, Qal+ helper T cell subset that sees both antigen and determinants on immunoglobulin, *J. Cell. Mol. Immunol.,* 1, 50, 1983.

169. Rohrer, J. W., Odermatt, B. O., and Lynch, R. G., Idiotype-specific suppression of MOPC-315 IgA secretion in vivo: reversible blockade of secretory myeloma cells by soluble mediators, *J. Immunol.,* 121, 1799, 1978.

170. Gebel, H. M., Hoover, R. G., and Lynch, R. H., Lymphocyte surface membrane immunoglobulin in myeloma. I. M315-bearing T lymphocytes in mice with MOPC-315, *J. Immunol.,* 123, 1110, 1979.

171. Hoover, R. G. and Lynch, R. G., Lymphocyte surface membrane immunoglobulin in myeloma. II. T cells with IgA-Fc receptors are markedly increased in mice with IgA plasmacytomas, *J. Immunol.,* 125, 1280, 1980.

172. Hoover, R. G., Dieckgraefe, B. K., and Lynch, R. G., T cells with Fc receptors for IgA: induction of Tα cells in vivo and in vitro by purified IgA, *J. Immunol.,* 127, 1560, 1981.

173. Hoover, R. G., Dieckgraefe, B. K., Lake, J., Kemp, J. D., and Lynch, R. G., Lymphocyte surface membrane immunoglobulin in myeloma. III. IgA plasmacytomas induce large numbers of circulating adult-thymectomy-sensitive, −+, Lyt-1−2+ lymphocytes with IgA-Fc receptors, *J. Immunol.,* 129, 2329, 1982.

174. Milburn, G. L. and Lynch, R. G., Immunoregulation of murine myeloma *in vitro.* II. Suppression of MOPC-315 immunoglobulin secretion and synthesis by idiotype-specific suppressor T cells, *J. Exp. Med.,* 155, 852, 1982.

175. Lynch, R. G. and Milburn, G. L., Id315-specific T cells that suppress MOPC-315 IgA synthesis recognize a VH315 idiotope, *Fed. Proc. Fed. Am. Soc. Exp. Biol.,* 42, 688, 1983.

176. Parslow, T. G., Milburn, G. L., Lynch, R. G., and Granner, D. K., Suppressor T cell action inhibits the expression of an excluded immunoglobulin gene, *Science,* 220, 1389, 1983.

177. Milburn, G. L., Parslow, T. G., Goldenberg, C., Granner, D. K., and Lynch, R. G., Idiotype-specific T cell suppression of light chain mRNA expression in MOPC 315 cells is accompanied by a posttranscriptional inhibition of heavy chain expression, *J. Mol. Cell Immunol.,* 1, 115, 1984.

178. Cory, S., Adams, J. M., and Kemp, D. J., Somatic rearrangements forming active immunoglobulin μ genes in B and T lymphoid cell lines, *Proc. Natl. Acad. Sci. U.S.A.,* 77, 4943, 1980.

179. Dudley, J. and Risser, R., Amplification and novel locations of endogenous mouse mammary tumor virus genomes in mouse T-cell lymphomas, *J. Virol.,* 49, 92, 1984.

180. Haughton, G. and Arnold, L. W., Manuscript in preparation.

181. Bishop, G. A. and Haughton, G., H-2 control of expression of an idiotype shared by normal B cells and a B-cell lymphoma, *Immunogenetics,* 21, 355, 1985.

Chapter 7

MEMORY B CELL ACTIVATION, DIFFERENTIATION, AND ISOTYPE SWITCHING

Thomas L. Feldbush, Laura L. Stunz, and David E. Lafrenz

TABLE OF CONTENTS

I. INTRODUCTION

Memory B lymphocytes are critical for long-term immunity, but we know very little about their generation and the maturation processes that must occur prior to their appearance as the final guardians against antigenic insult. Long-term immunity is probably a very dynamic process, involving multiple molecular and cellular components, rather than a static, "on-call" type of arrangement. With this in mind one can pose many questions regarding the induction and maintenance of the memory B lymphocytes. For example, what is the source of memory cells and what are the cellular and molecular interactions required for their generation? Are there intermediate stages of maturation before fully mature memory cells are formed? What are the physical and functional properties of immature and mature memory B cells? Is there only one type of mature memory cell or are there multiple parallel lines of maturation leading to the generation of "specialized" memory cell subsets?

In the following pages we shall attempt to answer some of these questions by reviewing our current understanding of: (1) the steps involved in the generation of memory B cells, (2) the functional B cell subsets, (3) T cell requirements for memory B cell induction, (4) factors regulating the induction and expression of memory B cells, and (5) the clonal expansion of the memory B cell population.

II. HISTORICAL PERSPECTIVE ON MEMORY B CELL DEVELOPMENT

A rather extensive body of information on the site and kinetics of memory B cell development has accumulated. The memory B cells were originally characterized as predominantly long-lived small lymphocytes[1-8] which arise within the follicular region[9-11] of lymphoid tissue draining the site of antigen injection.[12-14] These memory cells are the progeny of a precommitted lymphocyte pool not involved in primary antibody formation[15-18] and are continually recruited by available antigen[19-23] from both the recirculating population and the tissue-fixed cells.[24] Once generated, the memory cells may either become recirculating cells or perhaps remain sessile within the draining node.[2,4,12,13,20,23,25,27]

Following immunization with particulate antigen, memory T cells can be seen within 2 to 3 days, while memory B cells are detectable by day 4.[22,28] Soluble antigen priming is slightly slower with memory T cells appearing on days 3 to 4 and memory B cells by day 8.[14] Memory B cells begin leaving the draining lymph node by 11 days after immunization[8] with some of these cells subsequently homing into distal nodes resulting in systemic immunity.[23,27,29,30]

After generation, the memory cells acquire the ability to recirculate and appear to enter a quiescent phase in which they can remain essentially unchanged for long periods of time.[8,31-35] Upon restimulation by antigen these cells proliferate and differentiate into plasma cells or they proliferate to expand the memory cell pool[33,36-41] with the available antigen concentration apparently regulating which pathway predominates.[36]

Butcher et al.[42,43] and Reichert et al.[44] described a large B cell which develops in the germinal centers of Peyer's patches and binds high amounts of peanut agglutinin (PNA^{hi}). This PNA^{hi} population is distinct from other B cells, plasma cells, and most T cells which bind low levels of PNA (PNA^{low}). Using both immunofluorescent staining and panning, these PNA^{hi} cells were found to be negative for membrane IgD (mIgD), weakly positive for mIgM or mIgG (10 to 30%), but strongly positive for surface IA-encoded determinants. Coico et al.[45] isolated PNA^{hi} B cells from murine spleens at various times after immunization and tested their ability to adoptively transfer secondary responses. Ten days after immunization the PNA^{hi} B cells were enriched for memory cells, while 14 weeks later only PNA^{low} memory cells were found. This PNA receptor, found on immature, murine memory cells, is the first unique marker ever associated with the memory cell population.

III. MEMORY B CELL SUBSETS

As suggested by the data described above, most memory cell subsets have been defined on the basis of physical criteria. These physically defined subsets can be ordered into a hypothetical maturation sequence which can in turn be related to models of memory B cell generation. Table 1 presents such a maturation sequence and outlines three broad categories of memory cells: two populations of immature cells and a third population of mature cells. The latter subset is still quite heterogenous as defined by the expression of surface Ig.

A. Immature Memory Cells

Strober[46] was the first to describe a large, apparently immature, memory cell subpopulation found in the thoracic duct lymph (TDL) 4 to 8 weeks following priming. This cell population was four to five times more active on a per cell basis in adoptive transfer experiments than the small memory cell population. Sixteen weeks after priming the response within the large cell fraction waned while the small cell fraction maintained its activity. Using tritiated thymidine suicide, Strober demonstrated that these large cells were short-lived, since a 48-hr exposure to (^{3}H) thymidine before transfer to adoptive recipients markedly reduced the subsequent response to antigen. The small cell population was resistant to this treatment.

Hobbs and Feldbush[47] confirmed the appearance of large memory cells in the TDL early after priming. In addition, they analyzed the size distribution of the memory population within the draining lymph node from day 5 to day 77 after immunization. At 5 days the DNP memory population was restricted to the large cell fraction and with time (77 days) shifted, leaving only 10% of the activity in the large cell fraction and 46% residing in the small cell fraction. Hobbs and Feldbush[47] clearly demonstrated that the shift in the size distribution was the result of a memory cell maturation process rather than the delayed appearance of an unrelated small memory cell. This was demonstrated by transferring only the large cell fraction to an intermediate host. After 7 or 11 days TDL was collected, the cells separated by $1 \times g$ velocity sedimentation, and the large, medium, and small cell fractions transferred to final irradiated hosts. The anti-DNP responses of these final hosts indicated that after 7 days 53% of the memory activity was found in the medium fraction, and after 11 days the majority of the memory response was in the small cell fraction. Although this study did not directly assess the life span of these large cells, other investigators have documented that in $1 \times g$ velocity separation the large cell fraction is composed primarily of cells in S phase or immediately post-S phase.[48] Together, these studies indicate that the initial memory cell is a large, proliferating blast which matures into the classical small memory B lymphocyte.

As shown in Table 1, the large memory cell population can be divided into two subsets based upon their expression of mIgD. Lafrenz et al.[49] and Herzenberg et al.[50] have found that most or all immature memory cells are mIgD+ when sorted by FACS and tested in adoptive recipients. As described above, Butcher et al.[42,43] and Reichert et al.[44] have found that early after immunization a large immature memory cell population appears which lacks mIgD and binds PNA. It is possible that these mIgD− large cells are rapidly dividing precursors whose lack of surface delta reflects their proliferative state. Kanowith-Klein et al.[51] and Ashman[52] have described such a phenomenon where 5 days after in vitro stimulation, B cells lose mIgD and display a cyclical modu-

Table 1
MEMORY B CELL SUBSETS

Characteristic	Immature cells ————→Mature cells		
Size	Large —→Large	————→	Small
Surface Ig expression	sIgD⁻	sIgD+	sIgD+
	sIgM±	sIgM+	sIgD⁻
			sIgM+
			sIgM⁻ sIgG+
			sIgM⁻ sIgG⁻
			sIgG+
			sIgG⁻
PNA binding	Positive	Negative	Negative
Mel-14 binding	Negative	Negative (?)	Positive
Ia expression	High	Low (?)	Low
Complement receptor	Negative	Negative	50% Positive
Recirculation	No	No	Yes

lation of surface IgM and IgG. By FACS analysis and immunoprecipitation, other investigators have confirmed this loss of mIgD following in vitro activation of B cells.[53-56] The weak surface expression of IgM and IgG by PNA[hi] cells and the increased expression of Ia antigen[57] would be consistent with this hypothesis. It is even possible that the mIgD− cells are precursors of the mIgD+ cells, but this hypothesis must await formal proof.

In addition to the mIgD-positive and-negative large memory cells, evidence for an early mIgM+ population has also been reported. Strober[46] showed that the large memory cells were mIgM+ and, while not focusing on this point, the results of Teale et al.[58] and Zan-Bar et al.[59] would tend to agree that there is a significant portion of mIgM+ memory cells. On the basis of this information we would propose that the mIgD−, PNA[hi], weak mIgM, and mIgG-positive subset precedes in appearance the mIgD+, PNA[low], mIgM-positive subsets and may be their precursors (Table 1).

Regardless of their surface phenotype, all large memory cells lack complement receptors (CR).[60] Lymphocyte suspensions were prepared from antigen-draining lymph nodes at 2 weeks, 4 weeks, and 8 months after immunization with DNP-BGG. The cells were then separated by EAC rosetting and transferred to carrier-primed adoptive recipients. Total antibody responses attributable to CR+ and CR− cells were calculated by determining the amount of immunoglobulin secreted per cell transferred. At 2 weeks post-immunization 93% of the memory cells lacked CR, while at 4 weeks 71% were CR− and at 8 months the response was distributed equally among CR+ and CR− memory cells. Since both the mIgD− and mIgD+ populations and the PNA[hi] and PNA[low] populations exist at 2 weeks post-immunization and few, if any, CR+ memory cells are seen at this time, both immature populations are designated as CR− (Table 1).

Mason and Gowans[61] were the first to suggest that large memory cells arise in antigen-draining lymph nodes and these cells can circulate through the thoracic duct to other lymphoid tissues, but are unable to recirculate. This hypothesis was tested directly[62] and it was shown that large memory cells could be recovered from lymph nodes draining the site of antigen injection, but could not be recovered from the thoracic duct lymph following adoptive transfer.[47] Furthermore, as described above, Hobbs and Feldbush[47] were able to show that the large memory cells differentiate into smaller, more mature cells in the process of becoming a recirculating population. Butcher et al.[42,43] and Reichert et al.[44] have confirmed these observations using a monoclonal antibody to a putative receptor for the post-capillary venule. This monoclonal antibody is called Mel-14 and fails to bind to the PNA[hi] memory B cell subset.[44] Since none of the immature, large cells seem capable of recirculating, we have labeled both of the immature cell subsets as nonrecirculating and probably Mel-14 negative, but this has not been formally demonstrated (Table 1).

Since the PNA[hi] large cell subset is strongly Ia positive while the other B cells are much lower in surface Ia antigen expression, we propose that the putative immature mIgD+, PNA[low] population is also low in surface Ia (Table 1).

B. Mature Memory Cells

The phenotypic characteristics of the small, mature memory cells are less well characterized. As shown in Table 1, it now seems clear that the mature memory B cells are small,[1-8] quiescent,[1-8] PNA[low],[42-45] Ia[low],[42-44] recirculating,[2] and may either display or lack complement receptors.[60] Since all these cells are recirculating,[2] we have assumed that they are Mel-14 positive. The controversy surrounds the surface Ig (IgM, IgG, and IgD) displayed on these populations and it is these markers which will ultimately describe the number of memory B cell subsets.

Herzenberg et al.[50] and Black et al.[63] have proposed that early after priming (<10 weeks) there exist both mIgD+ and mIgD− memory cell subsets, but at later time points the mIgD− (presumably mIgG+) cells predominate. In addition, the mIgD+ population adoptively transferred appreciably lower affinity responses than did mIgD− memory cells. In contrast, Lafrenz et al.[49] examined several permutations of priming and transfer, but were never able to demonstrate significant differences in the affinity of the response restored by mIgD+ and mIgD− memory cell subsets. Two main points may account for the observed differences. Herzenberg et al.[50] and Black et al.[63] used the transfer of Ig-1b cells into Ig-1a hosts and Bosma et al.[64] have demonstrated a 1b allotype-specific natural suppressor present in 1a allotype hosts. Herzenberg et al.[50] made antibody determinations on day 7 post-challenge when, according to Bosma et al.,[64] suppressor cells would still be functional. However, on day 14 the suppressor T cells were no longer functional, and Herzenberg et al.[50] reported that at this time mIgD+ cell recipients demonstrated high affinity antibody secretion. Lafrenz et al.[49] utilized a protocol in which all classes of immunoglobulin were included in affinity determinations. In the other studies[50,63] only IgG$_{2a}$ and IgG$_1$ antibody responses were examined. One must also acknowledge that the differences in mouse strains used in these two studies[49,50,63] could affect the percentages and characteristics of mIgD+ and mIgD− memory cells.

Another point which remains controversial is the identity of the subpopulation of memory cells responsible for propagation of the memory response. Zan-Bar et al.[65] reported that only intermediate hosts reconstituted with mIgD+ cells were able to transfer memory responses to final hosts. However, Black et al.[66] demonstrated that both mIgD+ and mIgD− recipients were able to transfer memory responses to final hosts. In addition, Black et al.[66] showed that the cells from mIgD+ recipients, when transferred to final hosts, were mIgD−. This is consistent with their maturation scheme wherein mIgD− cells represent the most mature memory cell population. Final analysis of memory cell propagation awaits an in vitro system wherein expansion can be analyzed without host influence and surface isotypes can be continually monitored.

C. Surface Isotype and Immunoglobulin Class Commitment of Memory B Lymphocytes

In addition to the heterogeneity in the expression of mIgD, there is also great variation in the expression of mIgM and mIgG on memory B cells (Table 1). Using FACS separation, Okumura et al.[67] examined the ability of cells bearing IgG$_1$, IgG$_2$, and Ig-1b allotype to adoptively transfer a secondary immune response from 2-month DNP-

KLH-primed mice donors and found that the majority of the response consisted of the same immunoglobulin class as displayed on the transferred cells. In similar experiments, Mason and Gowans[61] and Mason[68] evaluated the surface isotype of 4- to 6-week DNP-primed thoracic duct cells and found that only the mIgG+, and not IgM+ cells, gave rise to indirect PFC responses in the spleens of adoptive recipients. In addition, the mIgG+ cells were shown to exist in the small cell fraction as determined by $1 \times g$ velocity sedimentation. In support of these studies, Coffman and Cohn[69] demonstrated that pretreatment of primed spleen cells with anti-IgG, serum protected these cells from antigen suicide by heavily iodinated TNP-KLH and allowed them to respond in adoptive transfer.

In negative selection experiments, Yuan et al.[70] treated SRBC-primed cells with anti-IgM and/or anti-IgG plus complement before adoptive transfer. Anti-IgG treatment decreased the adoptive secondary IgG response an average of 70%, which supported previous studies showing IgG memory being mediated by IgG-bearing cells. When memory cells were treated with anti-IgM alone, the adoptive response was not inhibited. However, when cells were treated with anti-IgM in addition to anti-IgG, the adoptive IgM response was ablated. From this, they concluded that cells responsible for the secondary IgM response have a lower density of IgM than cells which are responsible for the transfer of the primary response and, in addition, these secondary IgM cells bear small amounts of IgG on their surface. This was the first description of memory cells demonstrating dual expression of surface IgM and IgG.

In experiments described above, Strober et al.[46] showed that the large memory cell population was mIgM+. In the same experiments the small memory cells were shown to bear mIgG. It is important to note that Strober[46] found both populations (mIgM+ large and mIgG+ small) secreted IgG antibody. This was the first evidence that mIgM-positive memory cells were able to generate IgG-secreting cells which contradicts previous studies in which IgG memory was shown to be mediated solely by mIgG+ cells. However, Strober's data must be taken with some reservation since the presence of IgM antibody was inferred by the ability of 2-mercaptoethanol (2-ME) to reduce antigen binding in a Farr assay. No formal proof for this assumption was presented. These results may address an enigma which has concerned several investigators for years, i.e., if IgG memory is indeed mediated only by IgG-bearing memory cells, how does such a small cellular repertoire account for such a major portion of the immune response? Additional evidence that IgG memory could be mediated by IgM-bearing cells was reported by Abney et al.[71] They demonstrated that a memory cell clone with receptors of the IgM isotype produced IgG$_1$ or IgG$_2$ antibody upon reexposure to antigen. In addition, Pernis et al.[72] were able to demonstrate primed rabbit cells which bore surface IgM but cytoplasmic IgG after antigen challenge.

Zan-Bar et al.[73] expanded on these studies by using FACS separation of primed cells. Spleen cells from mice primed with DNP-BSA 8 to 12 weeks earlier were stained for surface IgM, IgD, or IgG and separated into dull and bright fractions. Enriched cells were adoptively transferred along with an excess of carrier primed cells to irradiated recipients, and antibody (2-ME sensitive and 2-ME resistant) was determined 7 and 9 days after challenge using the Farr assay. When cells bearing surface IgM or IgD were transferred, a significant IgM response was present on day 7, but the response became predominantly IgG by day 9. The transfer of IgG-bearing cells resulted in only IgG antibodies at the time points studied. Of further interest was the observation that removal of IgG-bearing cells resulted in an increase in the IgM response as compared to unfractionated cells. Apparently, the presence of an IgG response results in suppression of the adoptive secondary IgM response. Further studies by Zan-Bar et al.[59] examined the adoptive secondary response of isotype-predominant cells using the experimental protocol outlined above. IgM-predominant cells were able to transfer an adoptive secondary IgM response on day 7 post-challenge, whereas the response was predominantly IgG by day 9. IgD- and IgG-predominant cells restored an IgG-adoptive secondary response. It is important to emphasize that memory cells appear to be able to secrete IgM only if IgM is expressed on the cell surface. In addition, the production of IgG by IgM-bearing memory cells is consistent with the previously discussed conversion of the large IgM memory cells to small long-lived IgG-bearing cells.[46,47]

The production of specific isotypes by primed cells has been measured in the splenic focus assay. The impetus for these studies was to examine two points: (1) to reassess the ability of primed cells to secrete antibody isotypes different than those expressed on the cell membrane and (2) to investigate whether gene deletion accompanies membrane isotype switching in primed cells. In studies by Lafrenz et al.[74,75] and Teale et al.[58] the isotype production of 2- to 3-month primed cells separated on the basis of surface IgM, IgD, and IgG expression were examined. Spleen cells were stained for surface immunoglobulin, separated into bright and dull cell fractions using the FACS, and adoptively transferred to irradiated (1300R) carrier-primed recipients. Spleens were removed from adoptive hosts 18 hr post-transfer, and fragments were prepared and placed in tissue culture and challenged. Culture supernatants were analyzed by radioimmunoassay for all immunoglobulin isotypes at three time points after challenge (days 9, 12, and 16). As can be seen in Table 2, the clonal precursor frequency of IgG+ and IgG predominant cells is greater than the clonal frequency of IgM-bearing cells. However, since the IgM-bearing cells were present in greater numbers they still constituted more than half the adoptive secondary response. Also, it can be seen that the frequency of IgD-bearing cells was less than that of cells not expressing IgD, but as with mIgM+ cells, the IgD-bearing cells constituted a larger proportion of the response. When the specific isotypes produced by clonal progeny were examined (Table 3), IgM-sorted cells gave rise to three distinct sets of clones: those which produced IgM only, those which produced IgM plus IgG, and those making IgG alone. All IgG sorted cells produced IgG, but some clones (22%) still produced IgM. Ninety-six percent of the IgG predominant cells (IgM−, IgG+) produced IgG in the absence of IgM. If this result is reproducible, then one can assume that cells which express IgG alone are restricted to isotype production other than IgM. One clone was observed to produce IgM plus IgA, which is characteristic of immature cells and may have been due to cytophilic IgG. Since some mIgG+ cells (not mIgG predominant) still secrete IgM antibody (22% of the clones), this means that a subset of the memory B cells probably bear both mIgM and mIgG.

Based upon current understanding of the genetic mechanisms controlling the dual expression of mIgM and mIgG, it becomes necessary to formally prove that IgG-bearing cells result from endogenously synthesized IgG. Primed spleen cells were stripped of surface immunoglobulin by capping using antiimmunoglobulin antibody followed by protease digestion.[76] This procedure removes almost 100% of surface immunoglobulin. Cells were then cultured overnight to allow mIg resynthesis, stained, sorted, and transferred to adoptive recipients and then studied in the splenic focus assay. As can be seen in Table 3, the vast majority of IgG-bearing cells (>90%) resulted in clones which produced IgM. There were no major differences in the isotype production observed with IgD+ or IgD− singly sorted memory cells, as both transferred IgM and IgG production. The IgM production by mIgD− cells could result from cells that were mIgM+, mIgM+ plus mIgG+, or only mIgG+ and thus would not have been removed by a single FACS sort.

Current concepts of immunoglobulin molecular biology suggest that isotype switching occurs by either sister chromatid exchange[77-79] or allelic deletion.[80-84] The finding that mIgG+, mIgM− cells produced only IgG is consistent with the allelic exclusion model. Furthermore, the cells were not restricted in isotype production as predicted by

Table 2

FREQUENCY OF DNP-SPECIFIC MEMORY B CELL
PRECURSORS IN B CELL SUBSETS SEPARATED ACCORDING
TO SURFACE ISOTYPE

Experimental series	Experimental group[a]	% Of spleen cells stained	Average frequency of DNA-specific clones per 10^6 cells transferred	Approximate % of total response
A	Unfractionated		9.0	100
	μ^+	40—45	7.5	56.7
	$\mu^-\gamma^+$	4—5	15.4	30.5
	$\mu^-\gamma^-$		1.8	12.5
	γ^+	8—10	26.2	34.5
	γ^-		5.0	65.4
B	Unfractionated		33.7	100
	δ^+	50—55	26.6	40
	δ^-		58.3	45

[a] Spleen cells from primed animals were separated into subpopulations based on cell surface isotype using the FACS. The isolated subpopulations were injected into carrier primed irradiated recipients. Eighteen hours post-transfer, spleens were taken for in vitro challenge in the splenic focus assay.

Table 3

ISOTYPES SECRETED BY MEMORY B CELL SUBSETS SEPARATED
ACCORDING TO SURFACE ISOTYPE

Experimental series	Experimental group	No. of clones analyzed	% Of clones not secreting IgG vs. those secreting IgG[a]		
			IgG⁻	IgG+	
			(IgM, IgM, and/or IgA)	(IgM & IgG±IgA)	(IgG±IgA)
A	Unfractionated	44	4.5	18.1	77.3
	μ^+	32	9.4	21.9	68.8
	γ^+	32	0	21.9	78.2
	γ^-	13	7.6	30.7	61.5
	$\mu^-\gamma^+$	23	4.3	0	95.7
	$\mu^-\gamma^-$	4	50	0	50
B	Stripped[b] un-fractionated	33	30.4	57.5	12.1
	Stripped γ^+	43	46.5	46.5	7.0
	Stripped γ^-	16	62.5	12.5	25.0
C	Unfractionated	27	22.2	29.6	48.1
	δ^+	16	6.3	50.0	43.8
	δ^-	35	13.3	22.8	62.8

[a] Anti-DNP specific clones obtained in the splenic focus assay were analyzed for specific isotypes using a solid phase radioimmunoassay.

[b] Spleen cells were stripped of surface immunoglobulin by treatment with antiimmunoglobulin and protease treatment. Cells were then allowed to resynthesize surface membrane components by overnight incubation in tissue culture.

Cooper and co-workers,[85-88] as clones could produce any isotype which is 3′ in gene order to the isotype expressed on the cell surface. To summarize, priming of animals results in an increased frequency of IgG-bearing cells and of clonal progeny which are more restricted to the production of isotype 3′ to IgM. A similar conclusion has been reached in the studies of Gearhart and Cebra.[89]

While the data from the splenic focus assay dealing with IgG-predominant cells are consistent with the allelic deletion model of class switch,[58,74,89] other evidence suggests this model may not be complete. Recent studies with normal B cells continuously stimulated with LPS plus DXS have demonstrated deletion of the IgM gene in IgG-producing cells on not one, but both, chromosomes.[90] A particular problem with the model is its inability to explain how memory cells can co-express both mIgM and mIgG and how these cells can still secrete IgM antibody. Yaoita et al.[91] have described the presence of mIgM+ mIgE+ cells in mice which have been infected with *N. brasiliensis,* but DNA analysis has demonstrated a lack of DNA rearrangement. Therefore, co-expression of IgM and IgE must be mediated by RNA splicing mechanism similar to that which has been demonstrated for the co-expression of IgM and IgD.[92-94] Delineation of the molecular mechanisms operable in memory cells awaits technology which will allow clonal expansion of antigen-specific cells to a high-enough frequency to allow DNA analysis.

IV. INDUCTION OF MEMORY

A. Precursor Population

The classical models of Sercarz and Coons[95] for the generation of memory envisage a linear or branched pathway in which primary AFC and memory cells arise from a common precursor as outlined below. In this model X = virgin precursor, Y = memory cell, and Z = AFC.

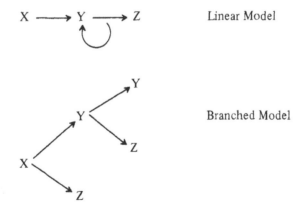

In the linear model *all* virgin cells must first pass through a memory cell stage *before* becoming AFC and thus the repertoire of clones (but not necessarily their quantity) would be the same in both primary and secondary responses. In the second model, the precursor cell could generate either a memory cell or an AFC depending upon the nature of the induction signals. Once formed, the memory cells would go through the same processes (linear or branched pathway) to generate plasma cells and to expand the memory cell clones. While there are anecdotal observations consistent with these models very little data exist that support either of them. Evidence exists which one may interpret to show that clonal expansion (and thus perhaps memory cell formation) and

AFC generation from a single set of precursors results from differential triggering. As will be discussed in greater detail later, the thymus-independent form of an antigen can stimulate the formation of AFC from primed cells, but cannot stimulate clonal expansion (Y $\rightarrow$ Z vs. Y $\rightarrow$ Y). A thymus-dependent form of the same antigen can stimulate both pathways. Elkins and Cambier[96] have described a B cell differentiation factor (BCDF) which stimulates the formation of AFC at the expense of cell proliferation. Thus, one could envision a situation in which antigen and lymphokines have stimulated the virgin B cell to proliferate and the level of BCDF would thus dictate the eventual outcome: high BCDF levels producing AFC and low levels allowing maximal memory B cell development. In support of the concept that proliferation and differentiation are separate and perhaps mutually exclusive events, Lafrenz et al.[97] have shown that treatment of the in vitro BCL_1 B cell tumor line with hydroxyurea dramatically increases LPS-induced differentiation while inhibiting further cell proliferation.

An alternate hypothesis pertaining to the generation of memory B cells would hold that the precursor cells for primary AFC and for memory are separate. In this model unique B cell precursors would give rise to parallel, but separate, lines of lymphocyte development. Goidl et al.[98] were the first to present evidence consistent with this hypothesis. Lethally irradiated mice were reconstituted with excess adult thymocytes and fetal liver cells as a source of immature B cells. Neonatal liver and adult spleen were used as a source of more mature B cells. The adoptive recipients were immunized with DNP-BGG and challenged 4 months later. PFC was measured after both immunizations and the heterogeneity of the responses was determined by hapten inhibition. From these results histograms were constructed which illustrated the distribution of low and high affinity AFC. It was found that fetal liver cells from 14- and 16-day fetuses reconstituted a primary immune response consisting of low affinity IgM antibody. However, following secondary challenge (at 4 months) the same animals displayed a high affinity IgG response. Neonatal and adult B reconstituted animals yielded high affinity AFC responses following both primary and secondary immunization. The authors concluded that B cells capable of generating high affinity memory cells develop before cells capable of generating the same type of primary AFC.

Szewczuk et al.[99] confirmed these studies using a variety of TD antigens and thus excluded an effect peculiar to DNP-BGG. Francus and Siskind[100] performed similar experiments between Igh allotype congenic mice in order to rule out a role of host B cells in the response. Again, they found that immature fetal liver cells will reconstitute mice for high affinity, high-magnitude secondary responses when either PFC or serum antibody is assayed. The same immature B cells yielded only a small, low affinity primary response.

While all of these studies suggest that the precursors for primary AFC and memory B cells are separate, there is a caveat inherent in the system. At the time of transfer, the B cells may be too immature to respond to antigenic stimulation. However, within a few days or weeks the cells could mature and become immunocompetent. At the same time the initial bolus of antigen would be metabolized leaving only low concentration available for lymphocyte stimulation. It is well known that low concentrations of antigens selectively stimulate the formation of memory B cells. Conclusions made from these studies may be compromised as a result of this artifact.

Evidence has accumulated in studies on primary and secondary responses to phosphocholine (PC) which also can be interpreted to show unique precursors for primary AFC and memory B cells. Wicker et al.[101] immunized CBA/N by BALB/c or DBA/2 F_1 male and female mice with PC conjugated to keyhole limpet hemocyanin (PC-KLH). The mice were either challenged directly or their cells transferred to adoptive recipients and challenged. Both serum and plaque-forming cell responses were then analyzed for the T15 idiotype in IgM and IgG antibodies specific for PC. Results from these studies suggest that the anti-PC response in normal mice (F_1 female) consists of two major antibody phenotypes and that these antibodies arise from different B cell subsets, the Lyb5+ and the Lyb5−. The Lyb5+ subset is present in female F_1 mice but lacking in male F_1 mice, and it produces TEPC 15 idiotype positive, PC inhibitable antibody found in the IgM, IgA, and IgG3 classes (called group I). The Lyb5− subset is present in both F_1 males and females and it produces TEPC 15 idiotype negative, phenylphosphocholine-specific (PPC) IgG_1 and IgG_2 antibodies (group II).[102] The group II antibodies are not inhibitable with PC but are with PPC and aminophenyl-phosphocholine. Group I antibodies were found to be dominant in the normal F_1 mice following primary immunization, while the group II antibodies were the predominant phenotype in secondary responses of both normal mice and the deficient male mice. This suggests that the predominant memory B cell population arises from a minor cell population which is either not expressed in the primary response or is expressed only weakly. These data support previous observations of Chang and Rittenberg[103] who showed a marked difference between the BALB/c anti-PC IgM and IgG memory response. They reported that greater than 90% of the primary IgM anti-PC response was T15+, while the IgG anti-PC secondary response consisted mostly of T15− antibody.

An alternate explanation for Wicker's results[101] would suggest that somatic mutation can occur upon immunoglobulin class switching and thus the T15+ precursor cell population could ultimately switch to the secretion of T15− antibodies. Chang et al.[104] tested this hypothesis by looking for the secretion of IgM, T15− group II antibodies in primed BALB/c mice. If somatic mutation accompanies class switching, no group II IgM antibodies should be seen. They found that early in the secondary response a large amount of group II IgM antibody was secreted and thus suggest that the group I (T15+) and group II (T15−) secondary antibodies arose from distinct germline genes. In addition, late in the secondary response the group II antibodies consisted mostly of IgG_1 immunoglobulins, probably resulting from class switching within the IgM-secreting population. As in the experiments of Wicker et al.,[101] this result argues that a large portion of the memory B cell population arises from a minor B cell subset which is only weakly expressed in the primary response.

Kenny et al.[105] used the splenic focus assay to examine Lyb5+ and Lyb5− precursor cells in primary and secondary responses. As anticipated, the Lyb5+ cells isolated from female (CBA/N × DBA/2) F_1 or (CBA/N × BALB/c) F_1 mice produced a T15+, IgM, and IgG primary immune response following immunization with PC conjugated hemocyanin (PC-Hy). However, the Lyb5− cells isolated from both normal (F_1 females) and defective (F_1, xid male) mice failed to respond in the in vitro splenic focus assay unless they came from a primed donor. Thus, while both normal and defective mice produce T15− responses when challenged *in situ*, the same cells do not respond in the splenic focus assay. On the basis of these observations Kenny et al.[105] proposed that the PC-specific Lyb5− virgin cell population lacks a receptor for BCDF and therefore fails to respond with differentiation in vitro. The Lyb5+ virgin cell population possesses these receptors and responds accordingly with the formation of AFC. However, the Lyb5− virgin cell population has a receptor for BCGF and, together with T cell help, can be stimulated to proliferate and differentiate into "primed" Lyb5− cells which now contain the full complement of BCGF and BCDF receptors. Again, these results suggest that the primary AFC and the memory cell have distinct precursors.

In summary, the only way to unequivocally demonstrate the existence of such unique B cell precursors is to isolate and characterize the cells.

B. The Role of T Cells in B Memory Generation

Investigators utilizing animals presumably depleted of T cells have provided evidence both for and against T cell involvement in the generation of memory cells. Adult animals which were thymectomized, lethally irradiated, and bone marrow reconstituted were not able to develop a secondary response to poly-L-(Tyr-Glu),[106] BGG,[107] or TNP-Brucella.[108] Adult thymectomy alone also resulted in a decreased ability to develop a secondary response to bovine serum albumin (BSA) or ovalbumin (OVA).[109] It may be argued that these experimental systems only show the inability to *express* a secondary response, since it has been shown that T cells are needed for the optimal differentiation of memory B cells to plasma cells.[110] This criticism was overcome by employing the adoptive transfer of cells from putative immunized donors to carrier-primed recipients. In this adoptive transfer system it has been reported that memory cells were either not generated or appeared in only low numbers in congenitally athymic mice[111,112] and in antilymphocyte serum-treated mice.[113] In allotype-suppressed animals which have been shown to have allotype-specific T suppressor cells and lack allotype-specific T_H cells, a memory response was shown upon adoptive transfer to carrier-primed recipients, but this response seemed to be due to only low-affinity memory cells.[114] Therefore, one would conclude that memory cells, at least high-affinity memory cells, may not develop in the absence of T cells. In contrast, other investigators using congenitally athymic mice showed memory responses when cells from these animals were challenged in vitro in the presence of T cells[115] or in the presence of supernatants from antigen-stimulated T cells.[116] Furthermore, it has been reported that adult thymectomized, lethally irradiated, and bone marrow-reconstituted animals develop a secondary response upon adoptive transfer[117] or when thymocytes were given at the time of antigen challenge.[118] Since these experiments utilized bone marrow cells as a B cell source, it is possible that low levels of T cells may have been present and this may account for the contradictory results reported.

Thymus-independent antigens offer another way to study the role of T cells in the development of memory B cells. Early experiments suggest that memory to these antigens does not exist,[119-121] while others claim to show limited memory, usually only of the IgM type following immunization with DNP-dextran,[122] DNP-levan,[123] and DNP-Ficoll.[124] With such a restricted response (IgM only) one must question the utility of such models. However, recently there have been some reports describing the development of IgG memory to TI antigens. For example, Schott and Merchant[125] found that congenitally athymic outbred mice, when immunized and challenged with DNP-Ficoll, demonstrated a strong IgM and a modest IgG secondary response. There were some unusual parameters to this secondary immune response to DNP-Ficoll: (1) the response was carrier specific and occurred in mice primed with Ficoll only, and (2) the qualitative and quantitative nature of the response in euthymic mice was highly strain dependent. Only IgM memory was seen in SJL/J, AKR/J, and C57/J mice, both IgM and IgG was seen in the C57BL/J mouse, and no memory was developed in C57BL/6N, C57BR/cdJ or A/HeJ mice.[126] Lite and Braley-Mullen[127] have found that polyvinyl-pyrrolidone (PVP) can prime for a secondary response, but only under very restricted conditions; when mice were given 2.5 ng PVP (very suboptimal immunogenic dose), secondary IgG responses could be elicited with a TD form of PVP, namely, PVP-horse red blood cells, but not with PVP itself. If large doses of PVP were used for priming (0.25 μg), a reasonable primary response was seen, but no memory development could be detected. While most of the studies listed above used TI-2 antigens to induce memory, Motta et al.[128] and Colle et al.[129] employed the TI-1 antigen trinitrophenylated lipopolysaccharide (TNP-LPS). C57BL/6 mice were immunized with an optimal immunogenic dose of TNP-LPS and the parameters of the antihapten response were obtained after challenge both in vivo and in vitro with homologous antigen. Compared to the primary response, the secondary response was both larger and contained more IgG antibody. The priming was relatively T independent since it could also be achieved in irradiated hosts reconstituted with spleen cells depleted of Thy-1-positive cells.

While these results obtained with TI antigens are encouraging, they still do not represent an ideal system for studying the role of T cells in B memory cell formation, since the responses are not reproducibly found, do not always display the classic isotype switch, and T cell influences have not been rigorously excluded from all experiments.

A better approach to this question would be the use of T cell-deficient animals reconstituted with graded numbers of mature T cells. If T cells are needed for the generation of memory cells, it would be predicted that the memory response would increase with increasing numbers of T cells participating in the response. This strategy circumvents the possible objection to previous experiments regarding residual T cells in T-deprived animals and allows for the evaluation of the role of T cells in the selection of high-affinity memory B cells. In primary responses both the affinity and heterogeneity of the antigen response increase as a function of time.[130-132] In the absence of T cells this maturation apparently does not occur.[133,134]

Lafrenz and Feldbush[135] have used this type of system to study the role of T cells in the development of memory B cells. Adult LBN rats were thymectomized, lethally irradiated 2 weeks later, and then reconstituted with 18-day fetal liver (a source of B cell precursors free of functional or phenotypically identifiable T cells). Four weeks later the animals were given graded doses of T cells and immunized with DNP-BGG. After an additional 4-week period, cells were recovered from these "primed" animals and transferred to carrier-primed irradiated host. This step ensures the presence of adequate T cell help for the *expression* of memory B cells. Serum antibody to DNP was measured after both primary and secondary immunization using the Farr technique, and the antigen binding capacity (ABC) was calculated. As shown in Table 4, little, if any, primary antibody formation was detected in animals receiving no exogenous T cells, and these same animals showed no significant memory B cell development. With increasing levels of T cell reconstitution both the primary and secondary responses increased, reaching levels equivalent to those seen in normal controls. When the average affinity of the secreted antibody was calculated (Table 4), a 29-fold increase was observed between primary and secondary antibody in the control group, and this is consistent with the maturation in affinity seen in normal animals. With no or low levels of T cell reconstitution (1×10^5) the affinity of the initial response was low, and little change was seen following challenge, suggesting that no maturation occurred. As the level of T cell reconstitution increased, so did the average association constant and the maturation of affinity. On the basis of all the results described above we would propose that T cells are probably needed for the induction of memory B cells, at least for those capable of secreting high affinity IgG antibody.

V. ACTIVATION OF VIRGIN AND MEMORY B CELLS

The activation of B lymphocytes has been the topic of several recent reviews.[136-141] In general, B cell activation is described as a multistep process, in which the cells are triggered under the influence of antigen-specific and nonspecific accessory cells and factors to proliferate and differentiate to AFC. The two processes of proliferation and differentiation are generally thought to be stimulated by distinct mechanisms.[96,97,142-148] Alternate pathways for activation may exist[148-150] and distinct cell populations may require different triggers.[148-156]

A. Polyclonal Activation

Early studies of B cell activation often employed lipopolysaccharide (LPS) as a B cell mitogen. In murine spleen cell cultures, this mitogen drives B cells to proliferate and secrete immunoglobulin in a polyclonal fashion.[157,158] Studies of human cells have generally employed other mitogens such as pokeweed mitogen (PWM) and *Staphylo-*

Table 4
ROLE OF T CELLS IN THE INDUCTION OF HIGH AFFINITY MEMORY B CELLS

Number[a] T cells given	Peak antibody[b] ABC (ng/ml)		Affinity[c] av Ko (L/M)		Ratio[d] Ko²/Ko¹
	Primary	Secondary	Primary	Secondary	
0	269	220	4.8×10^7	7.6×10^7	1.58
0.01×10^7	925	410	1.1×10^8	9.8×10^7	0.90
0.05×10^7	717	763	1.7×10^8	5.1×10^8	3.0
0.2×10^7	1320	940	1.5×10^8	7.6×10^8	5.07
1.0×10^7	2418	2024	1.3×10^8	1.9×10^9	14.6
Normal conc.	2242	1587	1.4×10^8	4.1×10^9	29.3

[a] Adult thymectomized irradiated rats were given graded doses of T cells, immunized with DNP-BGG, and the primary response determined. The animals were then sacrificed at 4 weeks after immunization, the cells transferred to carrier primed irradiated recipients and challenged. Secondary responses were then determined in the adoptive recipients.
[b] Average antigen binding capacity (ABC) was calculated as Ag/ml whole serum.
[c] Average association constant was determined by equilibrium dialysis.
[d] Ratio of average association constant following secondary challenge divided by the average association constant following primary response.

coccus aureus Cowan A strain bacteria (SAC).[159,160] Antiimmunoglobulin has also been used in both murine and human systems.[161-163]

Many of the early studies using polyclonal activators did not rigorously exclude T cells or macrophages from the B cell preparations and thus did not distinguish between mitogen effects upon B cells and contaminating accessory cells. More recent work has relied upon highly purified B cell populations in order to more directly evaluate the requirements for additional factors in activation. Stimulation of murine B cells in low density cultures with suboptimal doses of anti-μ revealed that the macrophage-derived factor interleukin 1 (IL-1) and T cell-derived B cell growth factors (BCGF) were required in order to elicit a proliferative response.[132,164] The requirement for BCGF has been confirmed by other groups using anti-μ and anti-δ stimulated human B cell proliferation and for the continued propagation of B cell blasts generated by SAC and PWM stimulation.[156,165-167] While anti-Ig appears to drive virtually all resting (G_0) B cells from normal or xid male mice to enter G_1 phase of the cell cycle, it is uncertain what percentage of B cells can be further driven into S phase in response to BCGF.[151,152]

In contrast to the results described for anti-Ig stimulated cells, Wetzel and Kettman[146,147] have demonstrated that individual B cells in culture can be driven to proliferate in the absence of T cells or macrophages. They determined that LPS could drive approximately 20% of cells to divide, and a second polyclonal activator, dextran sulfate (DXS), could stimulate division of 5% of cells. However, the combination of LPS plus DXS could trigger division of approximately 80% of the cells, provided a source of supportive fetal calf serum was used. This figure has been confirmed for antigen binding cells.[168] Under less supportive conditions, it has been shown that the addition of a supernatant from the monoclonal Dennert alloreactive T cell line, C.C3.11.75,[169] or from the macrophage-like cell line WEHI 3[146,147] can increase the proportion of B cells stimulated to clonal expansion by LPS plus DXS. Thus, the apparent direct action of LPS plus DXS is partially dependent upon exogenous growth factors.

Based on acridine orange staining and cytofluorometric analysis, it has been proposed that DXS may act by triggering cells to leave G_0, while LPS acts on a partially activated population.[146,147] This is consistent with the work of DeFranco et al.[151,152] who found that it was predominantly cells of intermediate density, perhaps activated in vivo, which responded to LPS with proliferation. A large body of work has been devoted to the examination of differences in the activation requirements of subpopulations of murine B lymphocytes which differ in their expression of the Lyb 5 marker.[140,152,153] Both populations exist in normal mice, while the CBA/N strain appears to lack the Lyb 5+ subset, due to an X-linked genetic defect. As described above, normal B cells will respond to anti-μ, with all of the cells passing from G_0 to G_1 and with optimal anti-μ concentrations, approximately 50%, proceeding into S phase. The B cells from xid-defective mice respond normally to LPS (approximately 20% are stimulated to enter S) but show an abortive response to anti-μ.[151,152] Stimulation with anti-μ triggers conversion from G_0 to G_1, but no progression into S is seen regardless of anti-μ concentrations.[151,152] A subset of these cells can, however, be stimulated to enter S phase by the addition of LPS to the anti-μ-stimulated cells.[152] Thus, it is clear that activation requirements differ for different B cell subpopulations. While the heterogeneity can be accounted for, at least in part, by variations in state of activation,[151,152,170] the results discussed above may also suggest that cell subsets can be triggered by identical mechanisms, but may differ in their requirements for a continued response.

Differentiation to Ig secretion appears to be under the control of antigen-nonspecific T cell-derived factors that are termed T cell replacing factors (TRF) or B cell differentiation factors (BCDF).[139-145,169,171-182] These factors generally act on cells that have been previously triggered out of a resting state by antigen or mitogen. However, exceptions in which resting cells may be directly activated to Ig secretion without proliferation have been described.[148-150] In the anti-μ driven system, a requirement for at least two TRFs has been shown.[139,178] In addition, Pure et al.[179] and Isakson et al.[180] have described distinct isotype-specific BCDFs. BCDFμ is capable of stimulating IgM secretion by BCL$_1$ cells, while BCDFγ appears to trigger IgG$_1$ secretion in LPS-stimulated cells by an mIgG− cell subset apparently precommitted to IgG$_1$ secretion.[181] BCDFs specific for induction of secretion of IgA have also been described.[182] Taken together, these observations reiterate the concept that B cell subsets differ in their responsiveness to regulation by antigen-nonspecific factors.

Studies of the requirements of polyclonally activated B memory cells have not been performed using protocols which allow easy comparison of their activation requirements to those of virgin cells. In fact, there is a limited amount of evidence that would suggest that memory cells may be refractory to activation with such mitogens as LPS[183-185] and SAC,[142] but rigorous demonstration of this point has not yet been made. Results supporting the unresponsiveness of primed cell populations to polyclonal activators have been obtained in this laboratory.[186] DXS plus LPS stimulation of DNP-primed, positively selected sIg+ lymph node cells, in the presence of a source of nonspecific T cell and macrophage factors, did not typically lead to the expression of IgG anti-DNP-PFC, while an IgM response similar to that obtained from an unprimed cell population was seen (Table 5). Antigen-specific responses were augmented by the addition of nonspecific factors, demonstrating that the cells were responsive. Tittle and Rittenberg[185] found that TNP-KLH-primed cells failed to respond to LPS stimulation with proliferation (measured by BUdR plus light suicide) or differentiation to AFC.

Zubler and Glasebrook[184] demonstrated that IgG antigen-specific AFC arose from primed B cells in the presence of LPS and a source of nonspecific T cell factors, provided that specific hapten was included in the culture. In the absence of hapten, little anti-TNP-specific IgG was seen, though LPS yielded a considerable polyclonal IgM response by the reverse plaque assay. The IgG antihapten response was also more de-

Table 5

POLYCLONAL ACTIVATION OF MEMORY CELLS

TNP-KLH-primed sIg+ cells[a]	Ag[b]	CAS[c]	DXS/LPS[d]	Anti-TNP-PFC/ 10⁶ cells	
				Direct	Indirect
+	+	0	0	0	2,528
+	+	+	0	375	8,400
+	0	+	+	29,950	0
+	+	+	+	3,878	7,325

[a] Four-week primed lymph node cells from Lewis × Brown Norway (LBN) rats, selected by anti-Ig panning.

[b] TNP-KLH 0.2 μg/mℓ.

[c] Supernatant from 24-hr Con A-stimulated spleen cells, 50—80% SAS cut.

[d] DXS plus LPS used at a final concentration of 10 μg/mℓ each.

pendent on T cell factors than was the polyclonal LPS-driven response. Snow et al.[187] demonstrated that, though TD antigens did not trigger ABC-enriched virgin B cells, they increased their susceptibility to LPS-stimulated proliferation. Here again, while antigen alone may have no overt effect upon B cells, it must exert an influence on the cells as they become more responsive to LPS stimulation. In some studies, this scheme may be complicated by regulatory pathways that selectively affect certain cell populations, though the use of T cell-free B cell populations should limit these variables.

B. Antigen-Specific Activation

The antigen-specific activation of virgin B lymphocytes has been studied by groups utilizing highly purified populations of antigen-binding cells.[145,168,187-191] Several of these studies have included the use of single cell assays and thus allow an improved evaluation of the role of antigen and accessory cell-derived factors in the antigen-specific response.[168,189-191]

The activation requirements for virgin B cells in TI antigen systems have been studied by several groups. There is general agreement that TI antigens can drive hapten-purified B cells to proliferate,[168,191] though some authors see a requirement for T cell factors in this process.[189,190] Endres et al.[176] have described an absolute dependence upon T cells for the generation of AFC by TI-1 and TI-2 antigens in rigorously T and MØ-depleted B cell cultures. Others using hapten-binding cell systems have described low but significant levels of AFC generation in the apparent absence of these factors.[168,191] Though isolated hapten-binding B cells appear to be activated to varying degrees by both TI-1 and TI-2 antigens, T-dependent antigens appear to be incapable of stimulating these highly purified B cells to enter the cell cycle[168] or enter into any proliferative or differentiative response.[144,145,187,188] This is true even when these T-dependent antigen-stimulated cells are supplied with sources of antigen-nonspecific T cell and macrophage factors.[144,145,168,187,188] It has been further demonstrated that at least some TNP-binding B cells could be induced to proliferate in the presence of primed T cells and appropriate carrier, with or without hapten present.[145,187] When linked recognition was supplied, proliferation was more vigorous, the B cells became enlarged, and they were driven to differentiate by nonspecific macrophage or T cell factors.[145] Of interest is the fact that cells that were driven to proliferate by primed T helper cells in the absence of linked recognition were unable to differentiate to AFC in response to nonspecific factors.[145] As discussed later, the phenomenon of antigen-specific B cell proliferation without subsequent differentiation may be relevant to memory cell propagation.

Several investigators have noted that virgin B cells are able to respond in vitro to certain T-dependent antigens in the absence of specifically primed T cell help if T cell-derived factors are provided.[141,171-177] The antigens used in these studies are characteristically particulate in nature and include heterologous erythrocyte-bound antigens. Cambier et al.[168] examined the ability of TNP-SRBC to drive TNP-binding cells into cell cycle as evaluated by acridine orange staining. Even in the presence of IL-1, IL-2 and TRF entry into cell cycle was not seen.

Jaworski et al.[144] and Shiozawa et al.[188] utilized antigen binding cells specific for chicken red blood cells of the B^2 allotype and demonstrated a strict requirement for stimultaneous binding of both antigen and an antigen-specific, MHC-restricted, T cell-derived helper factor to B cells in order to trigger proliferation. The studies of Andersson et al.,[192] using purified T_H factors, have also suggested a role for antigen-specific, Ia-restricted T help in the blastogenic response to SRBC. These results contast with the studies of Marrack and colleagues,[141,177] who have shown that rigorously T-depleted B cells could give a primary in vitro response to SRBC in the presence of supernatants containing IL-1, IL-2, and IL-X; the latter factor was contained in the supernatants of the FS6-14.13 cell line and may be gamma interferon.[193] An additional factor in the interpretation of these results lies in the fact that the classic particulate antigen systems rely on a B cell source that has been primed in vivo 4 to 7 days prior to initiation of culture.[141,171,172,177] Such treatment may supply the cells with their initial triggers and may be consistent with a role for MHC-restricted, Ag-specific T cells in the B cell response to particulate antigen.

Several groups have reported that the in vitro secondary response to soluble protein antigen can be independent of linked recognition.[172,174,175,194,195-197] This is in contrast to the primary response to such antigens, in which linked recognition seems to be strictly required.[141,172,175] In the classic system described by Hunig et al.,[174,175] DNP-KLH-primed cells were stimulated to secrete IgG anti-DNP in the presence of DNP-BGG and KLH, with both hapten and carrier required for the maximal response, though physical linkage was not required. Tada et al.[194] demonstrated a similar result with DNP-KLH-primed cells. In these experiments the antimouse brain plus C'-treated B cells were mixed with ovalbumin-primed T cells and challenged in vitro with DNP-KLH plus ovalbumin. This led to the production of IgG anti-DNP-PFC, despite the apparent absence of linked recongition. These studies, when contrasted to the experiments of Snow et al.[187] and Noelle et al.,[145] which demonstrate a strict requirement for linked recognition in the differentiation of TNP-binding cells, raise the possibility that memory B cells may be less stringent than virgin cells in their requirements for linked recognition.

The experiments, however, are not entirely convincing. The experiments by Hunig et al.[174,175] utilized B cells that had been recently boosted in vivo and thus had the opportunity to undergo certain activation events prior to initiation of culture. Julius and colleagues[154,198] showed that primed and recently boosted cells were predominantly large and responded to MHC-nonrestricted, unlinked T cell help while long-term primed, unboosted populations required MHC-restricted linked recognition for in vitro triggering.[154,198] Others have shown that 2-week DNP-BGG-primed cells could be driven to differentiate in vitro by irrelevant antigens, while 4- to 5-week primed cells only respond to specific antigen challenge (see below and Table 6).[155] Thus, the differences in activation requirements may reflect the activation state of the cells, independent of the fact that they are memory cells. Tada's experiments utilized a long-term primed population.[194] However, the in vitro challenge with hapten did not utilize a heterologous carrier. It is possible that residual T cells in the B cell population were sufficient to supply linked recognition and render the cells responsive to the TRF.

The studies of Takatsu and colleagues[195-197] used an antibody to the TRF acceptor

Table 6

ANTIGEN-MEDIATED EXPANSION OF B-MEMORY CELLS

Cell population	Day 0 challenge			Day 10 challenge		
	PFC/ 10^6	Precursor frequency	Burst size	PFC/ 10^6	Precursor frequency	Burst size
DNP-KLH-primed cells + OVA-primed, DNP-OVA challenged	15,300	$1.1/10^3$	119	264,300	$6.3/10^3$	166
DNP-KLH-primed DNP-Ficoll day 0 + OVA cells and DNP-OVA day 10	800	$1.4/10^5$	51	21,900	$7.1/10^4$	109
Large cell fraction 2-week DNP-KLH-primed + OVA-primed cells + DNA-OVA	80,000			230,000		
Small cell fraction 2-week DNP-KLH-primed + OVA-primed cells + DNP-OVA	10,000			15,000		
DNP-KLH-primed SRBC-challenged day 0 OVA-primed + DNP-OVA day 10	3,900			800		

and demonstrated that primed and recently boosted B cells bore this acceptor molecule. Further, the primed cells would differentiate to antibody secretion when supplied with antigen and a culture supernatant from the B151K12 T cell hybrid clone. This differs from the study of anti-μ-stimulated virgin cells by Nakanishi et al.,[178] which showed a requirement for at least two antigen nonspecific T cell-derived factors to drive differentiation to antibody secretion. The early-acting TRF in these studies was contained in B151K12 supernatant, while a late-acting factor was contained in a supernatant from the EL-4 thymoma line. Though a direct comparison of primed and unprimed cells was not made in Takatsu's study, it is intriguing to speculate that memory cells may lose the requirement for the later-acting TRF.[199] Whether they also lose the requirement for linked recognition and whether in vivo activation influences this change are questions that remain unanswered.

C. Requirements for Memory Cell Expansion

One of the distinctive features of memory B cells as compared to virgin cells is their capacity for antigen-driven clonal expansion, both in vivo[36,200,201] and in vitro.[155,202] In the in vitro system described by Brooks and Feldbush,[202] cells from primed animals could be propagated in the presence of antigen and carrier-primed T helper cells (Table 6). On day 10 of culture, the expanded cell population was rechallenged with antigen and the antigen-specific PFC response measured 7 to 13 days later. The intensity of the PFC response to the day 10 challenge, as compared to the response to the day 0 challenge, was used to assess memory cell clonal expansion. In some cases, limiting dilution analysis was used to measure the change in precursor frequencies from day 0 to day 10 of culture. The increase in number of responding cells after in vitro expansion was due both to proliferation of precursors and to increase in the number of PFC obtained per precursor (burst size). During the expansion of DNP-BGG-primed cells with DNP-OVA in the presence of OVA-primed T helper cells, a mean increase in precursor frequency of 16- to 67-fold has been observed.[202]

Several observations concerning the requirements for memory cell activation have been made using this system.[155,202] The responsive memory cell subset is a large, proliferating cell which predominates in primed cell populations 2 weeks after challenge. Small, more mature memory cells, which are also capable of producing a memory response in in vivo adoptive transfer systems, require an additional blastogenic signal in vivo before they will respond in the in vitro expansion system. This blastogenic signal can be induced by either specific antigen or adjuvant challenge. It is therefore possible that the responsive cell population in this system may be similar to the previously described direct progenitor population.[203] Virgin populations do not respond in this system, even if given in vivo adjuvant challenge, despite the presence of carrier-primed help.[155] Adherent accessory cells are also required for expansion.[155]

The antigen requirements for expansion have also been addressed.[155,202] Expansion does not occur in the absence of antigen or in the presence of irrelevant antigen such as SRBC. This is true despite the observation that SRBC can drive a subpopulation of primed cells to differentiate. DNP-Ficoll, a TI-2 antigen, can stimulate expansion of a subpopulation of the DNP memory cell pool, while only driving a minor portion of the cells to differentiate. These DNP-Ficoll challenged cells are then responsive to a day-10 TD antigen challenge.

The above observation is inconsistent with a memory cell subpopulation model proposed by Tittle and Rittenberg.[185] Using long-term TNP-KLH-primed mouse spleen cells, they showed that selective killing of TI-2 responsive B cells by antigen challenge in the presence of BUdR plus light could be achieved without affecting the TD-responsive population's ability to produce PFC. They also found that challenge with TI-1 antigens under the same conditions could eliminate both TI-2 and TD-responsive subpopulations. They concluded that B memory cells capable of differentiating to produce IgG anti-TNP-PFC are defined by only two functionally distinct subpopulations, and that TI-1 antigens can trigger division in both of these subpopulations. However, Brooks[155] clearly showed that TI-2 antigens are capable of propagating memory cells that have risen from the TD antigen-responsive subset. In an in vivo expansion system, Rennick et al.[200,201] showed that memory cells adoptively transferred from a tobacco mosaic virus protein (TMVP)-immune mouse could be triggered to both expand and differentiate by challenge with a TD form of a major TMVP epitope, while challenge with a TI-1 form of the epitope led to differentiation of a memory subset in the absence of clonal expansion. This subset appeared to overlap with the TI-1-responsive subset. In their system, no PFC were detected after a secondary challenge with a TI-2 form of the antigen, perhaps because the TMVP priming did not stimulate a TI-2-responsive subset, or perhaps because the TI-2 form of the antigen triggered expansion in the absence of detectable differentiation. Clearly, differences between the experimental systems described exist, and further study must precede the formation of conclusions regarding the differential ability of alternate antigenic forms to stimulate either distinct subpopulations of memory cells or cells in different stages of maturation.

Based upon the results of Brooks and Feldbush[155] using DNP-Ficoll, it can be proposed that clonal expansion of memory may occur without differentiation. Putting this observation together with current models of virgin B cell activation, in which separate signals lead to proliferation and differentiation, one could predict that clonal expansion may result from a condition where proliferation signals predominate. This would be consistent with the branched model of Sercarz and Coons[95] summarized above and identifies relative concentrations of lymphokines as a controlling factor. In order to test this hypothesis more directly, Thomas and Feldbush[204] have studied anti-Ig-induced memory clonal expansion (Table 7). In these studies DNP-primed B cells were cultured with anti-Ig alone or with OVA-primed T cells plus DNP-OVA. At 3 days of culture the anti-Ig-stimulated cells were harvested, mixed with fresh OVA-primed cells, and challenged with DNP-OVA. On day 10 of culture the antigen-stimulated population was harvested and rechallenged in a similar way. The level of memory cell clonal

Table 7

ANTIIMMUNOGLOBULIN-MEDIATED EXPANSION OF VIRGIN AND
MEMORY B CELLS

Cell population	Day 0 challenge (precursor frequency)	Delayed challenge	
		Precursor frequency	Proliferation ratio
Unprimed cells, DNP-OVA challenged	0.123	0.074[a]	0.8
Unprimed cells, anti-Ig challenged	0.123	0.076[b]	1.4
DNP-primed, DNP-OVA challenged	8.3	15.6[a]	7.8
DNP-primed, anti-Ig challenged	8.3	8.2[b]	1.3

[a]   Response to DNP-OVA on day 10 with fresh OVA-primed T help.
[b]   Response to DNP-OVA challenge on day 3 or 5 with fresh OVA-primed T help.

expansion was determined by comparing the precursor frequency found on day 0 of culture to that obtained after anti-Ig or antigen induction proliferation (Table 7). Using primed cells, the precursor frequency on day 0 was 8.3/10^4 cells. Following antigen expansion the precursor frequency on day 10 was 15.6/10^4 cells harvested and the increase in total B cells was 7.8-fold. Thus, at a minimum, there was a 15-fold increase in total DNP-specific memory cells. Following anti-Ig stimulation the precursor frequency per 10^4 cells harvested on day +3 stayed the same while the total cells recovered increased 1.3-fold. This produced a total increase in DNP-specific memory B cells of 1.3-fold. This figure seems low, but when one compares the unchallenged cells which fail to survive in vitro (25% recovery after 3 days culture), an apparent expansion was seen. Thus, we would conclude that anti-Ig stimulation leads to the in vitro proliferation of all B cells, including the DNP-specific memory cells, and supports the concept that a simple proliferation signal can lead to the propagation of memory.

REFERENCES

1. Gowans, J. L. and Uhr, J. W., The carriage of immunological memory by small lymphocytes in the rat, *J. Exp. Med.,* 124, 1017, 1966.
2. Gowans, J. L., Lymphocytes, *Harvey Lect.,* 64, 87, 1970.
3. Bosman, C. and Feldman, J., Cytology of immunologic memory: a morphological study of lymphoid cells during the ananmestic response, *J. Exp. Med.,* 128, 293, 1968.
4. Hunt, S. V., Ellis, S. T., and Gowans, J. L., The role of lymphocytes in antibody formation. IV. Carriage of immunological memory by lymphocyte fractions separated by velocity sedimentation and on glass bead columns, *Proc. R. Soc. London Ser. B:,* 182, 211, 1972.
5. Ellis, S. T. and Gowans, J. L., The role of lymphocytes in antibody formation. V. Transfer of immunological memory to tetanus toxoid; the origin of plasma cells from small lymphocytes, stimulation of memory cells in vitro and the persistence of memory after cell-transfer, *Proc. R. Soc. London Ser. B:,* 183, 125, 1973.
6. Mason, D. W. and Gowans, J. L., Subpopulations of B lymphocytes and the carriage of immunologic memory, *Ann. Inst. Pasteur,* 127C, 657, 1976.
7. Shortman, K., Fidler, J. M., Schlegel, R. A., Nossal, G. J. V., Howard, M., Lipp, J., and van Boehmer, H., Subpopulations of B lymphocytes: physical separation of functionally distinct stages of B cell differentiation, in *Contemporary Topics Immunobiology,* Vol. 4, Weigle, W. O., Ed., Plenum Press, New York, 1976, 1.
8. McGregor, D. D. and Mackaness, G. B., The properties of lymphocytes which carry immunologic memory of ∅X174, *J. Immunol.,* 114, 336, 1975.

9. Durkin, H. G. and Thorbecke, G. J., Relationship of germinal centers in lymphoid tissue to immunologic memory. V. The effect of prednisolone administered after the peak of the primary response, *J. Immunol.*, 106, 1079, 1971.

10. Durkin, H. G. and Thorbecke, G. J., Homing of immunologically committed lymph node cells to germinal centers in rabbits, *Nature (London) New Biol.*, 238, 53, 1972.

11. Grobler, P., Buerki, H., Cottier, H., Hess, M. W., and Stoner, R. D., Cellular bases for relative radioresistance of the antibody-forming system at advanced stages of the secondary response to tetanus toxoid in mice, *J. Immunol.*, 112, 2154, 1974.

12. Smith, J. B., Cunningham, A. J., Lafferty, K. J., and Morris, B., The role of the lymphatic system and lymphoid cells in the establishment of immunological memory, *Aust. J. Exp. Biol. Med. Sci.*, 48, 57, 1970.

13. Stavitsky, A. B. and Folds, J. D., The differential localization of antibody synthesis and of immunologic memory in lymph nodes draining and not draining the site of primary immunization with hemocyanin, *J. Immunol.*, 108, 152, 1972.

14. Mohr, R. and Krawinkel, U., Helper T cell kinetics and investigation of antigen receptor expression on early and memory T helper cells, *Immunology*, 31, 249, 1976.

15. Cunningham, A. J., Studies on the cellular basis of IgM immunological memory, *Immunology*, 16, 621, 1969.

16. Kappler, J. W., Hoffmann, M., and Dutton, R. W., Regulation of the immune response. I. Differential effect of passively administered antibody on the thymus-derived and bone marrow-derived lymphocytes, *J. Exp. Med.*, 134, 577, 1971.

17. Axelrad, M. A., Suppression of memory by passive immunization late in the primary response, *J. Exp. Med.*, 133, 857, 1971.

18. Safford, J. W. and Tokeida, S., Antibody mediated suppression of the immune response: effect on the development of immunologic memory, *J. Immunol.*, 107, 1213, 1971.

19. Steiner, L. A. and Eisen, H. N., Symposium on in vitro studies of the immune response, *Bacteriol. Rev.*, 30, 383, 1966.

20. Feldbush, T. L. and Gowans, J. L., Antigen modulation of the immune response. The effect of delayed challenge on the affinity of anti-DNP-BGG antibody produced in adoptive recipients, *J. Exp. Med.*, 134, 1453, 1971.

21. Hamaoka, T., Kitagawa, M., Matsuoka, Y., and Yamamura, Y., Antibody production in mice. I. The analysis of immunologic memory, *Immunology*, 17, 55, 1969.

22. Cunningham, A. J. and Sercarz, E. E., The asynchronous development of immunological memory in helper (T) and precursor (B) cell lines, *Eur. J. Immunol.*, 1, 413, 1971.

23. Greene, E. J., Tew, J. G., and Stavitsky, A. B., The differential localization of the in vitro spontaneous antibody and proliferative responses in lymphoid organs proximal and distal to the site of primary immunization, *Cell. Immunol.*, 18, 476, 1975.

24. Strober, S., Maturation of B lymphocytes in the rat. II. Subpopulations of virgin B lymphocytes in the spleen and thoracic duct lymph, *J. Immunol.*, 114, 877, 1975.

25. Phillips, M. E., Quagliata, F., and Thorbecke, G. J., Effect of thoracic duct drainage on the primary and secondary immune resonses in rats. Induction of immunological tolerance, *Immunology*, 22, 565, 1972.

26. Inchley, C. J., Black, S. J., and Mackay, E. A., Characteristics of immunological memory in mice. II. Resistance of non-recirculating memory cells to antigen-mediated suppression of the secondary antibody response, *Eur. J. Immunol.*, 5, 100, 1975.

27. Thorbecke, G. J. and Bell, M. K., The proliferative and anamnestic antibody response of rabbit lymphoid cells in vitro. II. Effect of passive antibody on immunologic memory in lymph nodes contralateral to the site of antigen injection, *J. Immunol.*, 111, 1043, 1973.

28. Black, S. J. and Inchley, J., Characteristics of immunological memory in mice. I. Separate early generation of cells mediating IgM and IgG memory to sheep erythrocytes, *J. Exp. Med.*, 140, 333, 1974.

29. Weissman, I. L., Peacock, M., and Eltringham, J. R., Regional lymph node irradiation: effect on local and distant generation of antibody forming cells, *J. Immunol.*, 110, 1300, 1973.

30. Jacobsen, E. B. and Thorbecke, G. J., The proliferative and anamnestic antibody response of rabbit lymphoid cells in vitro. I. Immunological memory in the lymph nodes draining and contralateral to the site of a primary antigen infection, *J. Exp. Med.*, 130, 287, 1969.

31. Fecsik, A. I., Butler, W. T., and Coons, A. H., Studies on antibody production. XI. Variation in the secondary response as a function of the length of the interval between two antigenic stimuli, *J. Exp. Med.*, 120, 1041, 1964.

32. Nettesheim, P., Makinodan, T., and Williams, M. L., Regenerative potential of immunocompetent cells. I. Lack of recovery of secondary antibody-forming potential after x-irradiation, *J. Immunol.*, 99, 150, 1967.

33. Feldbush, T. L., Antigen modulation of the immune response. The decline of immunological memory in the absence of continuing antigenic stimulation, *Cell. Immunol.*, 8, 435, 1973.

34. Strober, S., Immune function. Cell surface characteristics and maturation of B cell subpopulations, *Transplant. Rev.*, 24, 84, 1975.

35. Cramer, M. and Braun, D. G., Immunological memory: stable IgG patterns determine in vivo responsiveness at the clonal level, *Scand. J. Immunol.*, 4, 63, 1975.

36. Feldbush, T. L. and van der Hoven, A., Antigen modulation of the immune response. IV. Selective triggering of antibody production and memory cell proliferation, *Cell. Immunol.*, 25, 152, 1976.

37. Askonas, B. A., Cunningham, A. J., Kreth, H. W., Roelants, G. E., and Williamson, A. R., Amplification of B cell clones forming antibody to the 2,4-dinitrophenyl group, *Eur. J. Immunol.*, 2, 494, 1972.

38. Williamson, A. R. and Askonas, B. A., Senescence of an antibody-forming cell clone, *Nature (London)*, 238, 337, 1972.

39. McMichael, A. J. and Williamson, A. R., Clonal memory. I. Time-course of proliferation on B memory cells, *J. Exp. Med.*, 139, 1361, 1974.

40. Klaus, G. G. B. and Willcox, H. N. A., B cell tolerance induced by polymeric antigens. III. Dissociation of antibody formation and memory generation in tolerant mice, *Eur. J. Immunol.*, 5, 699, 1975.

41. Nakashima, I. and Kato, N., Maintenance and amplification of cell-associated immunological memory in in vivo culture system, *Cell Immunol.*, 20, 156, 1975.

42. Butcher, E. C., Reichert, R. A., Coffman, R. L., Nottenburg, C., and Weisman, I. L., Surface phenotype and migratory capability of Peyer's patch germinal center cells, *Adv. Exp. Med. Biol.*, 149, 765, 1982.

43. Butcher, E. C., Rouse, R. V., Coffman, R. L., Nottenburg, C. N., Hardy, R. R., and Weissman, I. L., Surface phenotype of Peyer's patch germinal center cells: implications for the role of germinal centers in B cell differentiation, *J. Immunol.*, 129, 2698, 1982.

44. Reichert, R. A., Gallatin, W. M., Weissman, I. L., and Butcher, E. C., Germinal center B cells lack homing receptors necessary for normal lymphocyte recirculation, *J. Exp. Med.*, 157, 813, 1983.

45. Coico, R. F., Bhogal, B. S., and Thorbecke, G. J., Relationship of germinal centers in lymphoid tissue to immunologic memory. VI. Transfer of B cell memory with lymph node cells fractionated according to their receptors for peanut agglutinin, *J. Immunol.*, 131, 2254, 1983.

46. Strober, S., Maturation of B lymphocytes in rats. III. Two subpopulations of memory B cells in the thoracic duct lymph differ by size, turnover rate, and surface immunoglobulin, *J. Immunol.*, 117, 1288, 1976.

47. Hobbs, M. V. and Feldbush, T. L., Antigen modulation of the immune response. VI. Rate of large memory cell appearance in lymph nodes and thoracic duct lymph, *Cell. Immunol.*, 50, 30, 1980.

48. Crabtree, G. R., Munick, A., and Smith, K. A., Glucocorticoids an lymphocytes. II. Cell cycle-dependent changes in glucocorticoid receptor content, *J. Immunol.*, 125, 13, 1980.

49. Lafrenz, D., Strober, S., and Vitetta, E., The relationship between surface immunoglobulin isotype and the immune function of murine B lymphocytes. V. High affinity secondary antibody responses are transferred by both IgD-positive and IgD-negative memory B cells, *J. Immunol.*, 127 867, 1981.

50. Herzenberg, L. A., Black, S. J., Tokuhisa, T., and Herzenberg, L. A., Memory B cells at successive stages of differentiation. Affinity maturation and the role of IgD receptors, *J. Exp. Med.*, 151, 1071, 1980.

51. Kanowith-Klein, S., Vitetta, E. S., and Ashman, R. F., The isotype cycle: successive changes in surface immunoglobulin classes expressed by the antigen-binding B-cell population during the primary in vivo immune response, *Cell. Immunol.*, 62, 377, 1981.

52. Ashman, R. F., The immunological role of antigen-binding cells, *Immunol. Today*, 3, 349, 1982.

53. Preudhomme, J. L., Loss of surface IgD by human B lymphocytes during polyclonal activation, *Eur. J. Immunol.*, 7, 191, 1977.

54. Pernis, B., Brouet, J. C., and Seligmann, M., IgD and IgM on the membrane of lymphoid cells in macroglobulinemia. Evidence for identity of membrane IgD and IgM antibody activity in a case with anti-IgG receptors, *Eur. J. Immunol.*, 4, 776, 1974.

55. Sitia, R., Abbott, J., and Hammerling, U., The ontogeny of B lymphocytes. V. Lipopolysaccharide-induced changes of IgD expression on murine B lymphocytes, *Eur. J. Immunol.*, 9, 859, 1979.

56. Bourgois, A., Kitajiama, K., Hunter, J. R., and Askonas, B. A., Surface immunoglobulins of lipopolysaccharide-stimulated spleen cells. The behavior of IgM, IgD and IgG, *Eur. J. Immunol.*, 7, 151, 1977.

57. Monroe, J. G. and Chambier, J. C., Level of mIa expression on mitogen-stimulated murine B lymphocytes is dependent on position in cell cycle, *J. Immunol.*, 130, 626, 1983.

58. Teale, J. M., Lafrenz, D., Klinman, N. R., and Strober, S., Immunoglobulin class commitment exhibited by B lymphocytes separated according to surface isotype, *J. Immunol.*, 126, 1952, 1981.

59. Zan-Bar, I., Vitetta, E. S., Assisi, F., and Strober, S., The relationship between surface immuno-globulin isotype and immune function of murine B lymphocytes. III. Expression of a single predominant isotype on primed and unprimed B cells, *J. Exp. Med.*, 147, 1374, 1978.

60. Feldbush, T. L., Separation of memory cell subpopulations by complement receptors: in vivo analysis, *Eur. J. Immunol.*, 10, 443, 1980.

61. Mason, D. W. and Gowans, J. L., Subpopulations of B lymphocytes and the carriage of immunologic memory, *Ann. Inst. Pasteur*, 127C, 657, 1976.

62. Feldbush, T. L. and Stewart, N., Antigen modulation of the immune response. V. Generation of large memory cells in antigen draining lymph nodes, *Cell. Immunol.*, 37, 336, 1978.

63. Black, S. J., van der Loo, W., Loken, M. R., and Herzenberg, L. A., Expression of IgD by murine lymphocytes. Loss of surface IgD indicates maturation of memory B cells, *J. Exp. Med.*, 147, 984, 1978.

64. Bosma, M. J., Bosma, G. C., and Owen, J. L., Prevention of immunoglobulin production by allotype-dependent T cells, *Eur. J. Immunol.*, 8, 562, 1978.

65. Zan-Bar, I., Strober, S., and Vitetta, E. S., The relationship between surface immunoglobulin isotype and immune function of murine B lymphocytes. IV. Role of IgD-bearing cells in the propagation of immunologic memory, *J. Immunol.*, 123, 925, 1979.

66. Black, S. J., Tokuhisa, T., Herzenberg, L. A., and Herzenberg, L. A., Memory B cells at successive stages of differentiation: expression of surface IgD and capacity for self renewal, *Eur. J. Immunol.*, 10, 846, 1980.

67. Okumura, K., Julius, M. H., Tsu, T., Herzenberg, L. A., and Herzenber, L. A., Demonstration that IgG memory is carried by IgG-bearing cells, *Eur. J. Immunol.*, 6, 467, 1976.

68. Mason, D. W., The class of surface immunoglobulin on cells carrying IgG memory in rat thoracic duct lymph: the size of the subpopulation mediating IgG memory, *J. Exp. Med.*, 143, 1122, 1976.

69. Coffman, R. L. and Cohn, M., The class of surface immunoglobulin on virgin and memory B lymphocytes, *J. Immunol.*, 118, 1806, 1977.

70. Yuan, D., Vitetta, E. S., and Kettman, J. R., Cell surface immunoglobulin. XX. Antibody responsiveness of subpopulations of B lymphocytes bearing different isotypes, *J. Exp. Med.*, 145, 1421, 1977.

71. Abney, E. R., Keeler, K. D., Parkhouse, R. M. E., and Willcox, H. N. A., Immunoglobulin M receptors on memory cells of immunoglobulin G antibody-forming cell clones, *J. Immunol.*, 6, 443, 1976.

72. Pernis, B., Forni, L., and Luzzati, A. L., Synthesis of multiple immunoglobulin classes by single lymphocytes, *Cold Spring Harbor Symp. Quant. Biol.*, 41, 175, 1976.

73. Zan-Bar, I., Strober, S., and Vitetta, E. S., The relationship between surface immunoglobulin isotype and immune function of murine B lymphocytes. I. Surface immunoglobulin isotypes on primed B cells in the spleen, *J. Exp. Med.*, 45, 1188, 1977.

74. Lafrenz, D., Strober, S., Teale, J. M., and Klinman, N. R., Relationship between surface immunoglobulin isotypes and secreted isotypes during B cell differentiation, in *B Lymphocytes in the Immune Response: Functional, Developmental, and Interactive Properties*, Klinman, N. R., Mosier, D. E., Scher, I., and Vitetta, E., Eds., Elsevier/North-Holland, New York, 1981, 377.

75. Lafrenz, D., Teale, J. M., and Strober, S., Role of IgD in immunological memory, *Ann. N.Y. Acad. Sci.*, 399, 375, 1982.

76. Lafrenz, D., Teale, J. M., Klinman, N. R., and Strober, S., IgG bearing cells retain the capacity to secrete IgM, submitted.

77. Obata, M., Kataoka, T., Nakai, S., Yamagishi, H., Takahashi, N., Yamawaki-Kataoka, Y., Nikaido, T., Schimizu, A., and Honjo, T., Structure of a rearranged γ_1 chain gene and its implication to immunoglobulin class-switch mechanism, *Proc. Natl. Acad. Sci. U.S.A.*, 78, 2437, 1981.

78. Honjo, T., Kataoka, T., Yoita, Y., Shimizu, A., Takahashi, N., Yamawaki-Kataoka, Y., Nikaido, T., Nakai, S., Obata, M., Kawakami, T., and Nishida, Y., Organization and reorganization of immunoglobulin heavy chain genes, *Cold Spring Harbor Symp. Quant. Biol.*, 45, 913, 1981.

79. Rabbitts, T. H., Hamlyn, P. H., Matthyssens, G., and Roe, B. A., The variability, arrangement and rearrangement of immunoglobulin genes, *Can. J. Biochem.*, 58, 176, 1980.

80. Honjo, T. and Kataoka, T., Organization of immunoglobulin heavy chain genes and allelic deletion model, *Proc. Natl. Acad. Sci. U.S.A.*, 75, 2140, 1978.

81. Coleclough, C., Cooper, D., and Perry, R. P., Rearrangement of immunoglobulin heavy chain genes during B-lymphocyte development as revealed by studies of mouse plasmacytoma cells, *Proc. Natl. Acad. Sci. U.S.A.*, 77, 1422, 1980.

82. Cory, S., Jackson, J., and Adams, J. M., Deletions in the constant region locus can account for switches in immunoglobulin heavy chain expression, *Nature (London)*, 285, 450, 1980.

83. Rabbitts, T. H., Forster, A., Dunnick, W., and Bentley, D. L., The role of gene deletion in the immunoglobulin heavy chain switch, *Nature (London)*, 283, 351, 1980.

84. Yaoita, Y. and Honjo, T., Deletion of immunoglobulin heavy chain genes from expressed allelic chromosome, *Nature (London)*, 286, 850, 1980.

85. Kincade, P. W., Lawton, A. R., Bockman, D. E., and Cooper, M. D., Suppression of immunoglobulin G synthesis as a result of antibody-mediated suppression of immunoglobulin M synthesis in chickens, *Proc. Natl. Acad. Sci. U.S.A.*, 69, 918, 1970.

86. Cooper, M. D., Lawton, A. R., and Kincade, P. W., A two-stage model for development of antibody-producing cells, *Clin. Exp. Immunol.*, 11, 143, 1972.

87. Kearney, J. F., Cooper, M. D., and Lawton, A. R., B cell differentiation induced by lipopolysaccarride. IV. Development of immunoglobulin class restriction in precursors of IgG-synthesizing cells, *J. Immunol.*, 117, 1567, 1976.

88. Abney, E. R., Cooper, M. D., Kearney, J. R., Lawton, A. R., and Parkhouse, R. M. E., Sequential expression of immunoglobulin on developing mouse B lymphocytes: a systematic survey that suggests a model for the generation of immunoglobulin isotype diversity, *J. Immunol.*, 120, 2041, 1978.

89. Gearhart, P. J. and Cebra, J. J., Most B cells that have switched surface immunoglobulin isotypes generate clones of cells that do not secrete IgM, *J. Immunol.*, 127, 1030, 1981.

90. Radbruch, A. and Sablitzky, F., Deletion of $C\mu$ genes in mouse B lymphocytes upon stimulation with LPS, *EMBO J.*, 2, 1929, 1983.

91. Yaoita, Y., Kumagai, Y., Okumura, K., and Honjo, T., Expression of lymphocyte surface IgE does not require switch recombination, *Nature (London)*, 297, 697, 1982.

92. Maki, R., Roeder, W., Traunecker, A., Sidman, C., Wabl, M., Raschke, W., and Tonegawa, S., The role of DNA rearrangement and alternative RNA processing in the expression of immunoglobulin, *Cell*, 24, 353, 1981.

93. Morre, K. W., Rogers, J., Hunkapiller, T., Early, P., Nottenburg, C., Weissman, I., Bazin, H., Wall, R., and Hood, L., Expression of IgD may use both DNA rearrangement and RNA splicing mechanisms, *Proc. Natl. Acad. Sci.*, 78, 1800, 1981.

94. Knapp, M. R., Liu, C.-P., Newell, N., Ward, R. B., Tucker, P. W., Strober, S., and Blattner, F., Simultaneous expression of immunoglobulin μ and δ heavy chains by a cloned B-cell lymphoma: a single copy of the V_H gene is shared by two adjacent C_H genes, *Proc. Natl. Acad. Sci. U.S.A.*, 79, 2996, 1982.

95. Sercarz, F. F. and Coons, A., in *Mechanisms of Immunologic Tolerance*, Hašek, M., Lengerova, A., Vojtiskova, M., Eds., Czechoslovak Academy of Science Press, Prague, 1962, 73.

96. Elkins, K. and Chambier, J. C., Constitutive production of a factor supporting B lymphocyte differentiation by a T cell hybridoma, *J. Immunol.*, 130, 1247, 1983.

97. Lafrenz, D., Koretz, S., Stratte, P. T., Ward, R. B., and Strober, S., LPS-induced differentiation of a murine B cell leukemia (BCL₁): changes in surface and secreted IgM, *J. Immunol.*, 129, 1329, 1982.

98. Goidl, E. A., Klass, J., and Siskind, G. W., Ontogeny of B lymphocyte function. II. Ability of endotoxin to increase the heterogeneity of affinity of the immune response of B lymphocytes from fetal mice, *J. Exp. Med.*, 143, 1503, 1976.

99. Szewczuk, M. R., Sherr, D. H., Cornacchia, A., Kim, Y. T., and Siskind, G. W., Ontogeny of B lymphocytes function. XI. The secondary response by neonatal and adult B cell populations to different T-dependent antigens, *J. Immunol.*, 122, 1294, 1979.

100. Francus, T. and Siskind, G. W., Ontogeny of B lymphocyte function. XII. Evidence that the ability to generate memory cells precedes the ability to produce antibody-secreting cells, *Cell. Immunol.*, 72, 77, 1982.

101. Wicker, L. S., Guelde, G., Scher, I., and Kenny, J. J., Antibodies from the Lyb5- B cell subset predominate in the secondary IgG response to phosphocholine, *J. Immunol.*, 129, 950, 1982.

102. Wicker, L. S., Guelde, G., Scher, I., and Kenny, J. J., The asymmetry in idiotype-isotype expression in the response to phosphocholine is due to divergence in the expressed repertoires of Lyb-5+ and Lyb-5- B cells, *J. Immunol.*, 131, 2468, 1983.

103. Chang, S. P. and Rittenberg, M. B., Immunologic memory to phosphorylcholine in vitro. I. Asymmetric expression of clonal dominance, *J. Immunol.*, 126, 975, 1981.

104. Chang, S. P., Brown, M., and Rittenberg, M. B., Immunologic memory to phosphorylcholine. III. IgM includes a fine specificity population distinct from TEPC 15, *J. Immunol.*, 129, 1559, 1982.

105. Kenny, J. J., Yaffe, L. J., Ahmed, A., and Metcalf, E. S., Contribution of Lyb5+ and Lyb5- B cells to the primary and secondary phosphocholine-specific antibody response, *J. Immunol.*, 130, 2574, 1983.

106. Mitchell, G. F., Grumet, F. C., and McDevitt, H. O., Genetic control of the immune response. The effect of thymectomy on the primary and secondary antibody response of mice to poly-L(Tyr,glu)-poly-D,L-ala-poly-L-lys, *J. Exp. Med.*, 135, 126, 1972.

107. Williams, R. M. and Waksman, B. H., Use of thymectomized, irradiated rats to study immunogenicity of bovine gamma globulin, *J. Immunol.*, 120, 925, 1969.

108. Romano, T. J. and Thorbecke, G. J., Thymus influence on the conversion of 19S and 7S antibody formation in the response to DNP-Brucella, *J. Immunol.*, 115, 332, 1975.

109. Simpson, E. and Cantor, H., Regulation of the immune response by subclasses of T lymphocytes. II. The effect of adult thymectomy upon humoral and cellular responses in mice, *Eur. J. Immunol.*, 5, 337, 1975.

110. Miller, J. F. A. P. and Sprent, J., Cell-to-cell interactions in the immune response. VI. Contributions of thymus-derived cells and antibody-forming cell precursors to immunological memory, *J. Exp. Med.*, 134, 66, 1971.

111. Schlegal, R. A., Antigen-initiated B lymphocyte differentiation. Kinetics and T lymphocyte dependence of the primary and secondary in vitro immune responses to the hapten 4-hydroxy-3-iodo-5-nitrophenyl acetic acid presented on the carrier polymerized bacterial flagellin, *Aust. J. Exp. Biol. Med. Sci.*, 52, 471, 1974.

112. Dresser, D. W. and Popham, A. M., The influence of T cells on the initiation and expression of immunological memory, *Immunology*, 38, 265, 1979.

113. Braley-Mullen, H., Secondary IgG responses to type III pneumococcal polysaccharide. III. T cell requirement for development of B memory cells, *Eur. J. Immunol.*, 7, 775, 1977.

114. Okamura, K., Metzler, C. M., Tsu, T. T., Herzenberg, L. A., and Herzenberg, L. A., Two stages of B-cell memory cell development with different T-cell requirements, *J. Exp. Med.*, 144, 345, 1976.

115. Diamantstein, T. and Blitstein-Willinger, E., T cell independent development of B memory cells, *Eur. J. Immunol.*, 4, 830, 1974.

116. Schrader, J. W., The role of T cells in IgG production: thymus dependent antigens induce B-cell memory in the absence of T cells, *J. Immunol.*, 114, 1665, 1975.

117. Shinohara, N. and Tada, T., Hapten-specific IgM antibody responses in mice against a thymus-independent antigen (DNP-Salmonella), *Int. Arch. Allergy Appl. Immunol.*, 47, 762, 1974.

118. Roelants, G. E. and Askonas, B. A., Immunological B memory in thymus deprived mice, *Nature (London) New Biol.*, 239, 63, 1972.

119. Snippe, H., Nab, J., and van Eyk, R. V. W., In vitro stimulation of spleen cells of the mouse by DNP-carrier complexes, *Immunology*, 27, 761, 1974.

120. Klaus, G. G. B. and Humphrey, J. H., The immunologic properties of hapten coupled to thymus-independent carrier molecules. I. The characteristics of the immune response to dinitro-phenyl-lysine-substituted pneumococcal polysaccharide and levan, *Eur. J. Immunol.*, 4, 370, 1974.

121. Miranda, J. J., Studies on immunological paralysis. IX. The immunogenicity and tolerogenicity of levan in mice, *Immunology*, 23, 829, 1972.

122. Rude, E., Wrede, J., and Gundelach, M. L., Production of IgG antibodies and enhanced responses of nude mice to DNP-AE-dextran, *J. Immunol.*, 116, 527, 1976.

123. del Guercio, P., Thobie, N., and Poirier, M. F., IgM anamestic immune response to the haptenic determinant DNP on a thymus independent carrier, *J. Immunol.*, 112, 427, 1974.

124. Sharon, R., McMaster, P., Kask, A., Owens, J., and Paul, W., DNP-lys-Ficoll: a T independent antigen which elicits both IgM and IgG anti-DNP antibody-secreting cells, *J. Immunol.*, 114, 1585, 1975.

125. Schott, C. F. and Merchant, B., Carrier-specific immune memory to a thymus-independent antigen in congenitally athymic mice, *J. Immunol.*, 122, 1710, 1979.

126. Schott, C. F., Lizzio, E. F., Inman, J. K., and Merchant, B., Immune memory to a nonmitogenic, thymic independent antigen in mice: variation among inbred strains and possible relationship to oncogenesis, *J. Immunol.*, 127, 139, 1981.

127. Lite, H. S. and Braley-Mullen, H., Induction of IgG memory responses with polyvinylpyrrolidone (PVP) is antigen dose dependent, *J. Immunol.*, 126, 928, 1981.

128. Motta, I., Portnoi, D., and Truffa-Bachi, P., Induction and differentiation of B memory cells by a thymus-independent antigen, trinitrophenylated lipopolysaccharide, *Cell. Immunol.*, 57, 327, 1981.

129. Colle, J.-H., Motta, I., and Truffa-Bachi, P., Generation of immune memory by haptenated derivatives of thymus-independent antigens in C57BL/6 mice. I. The differentiation of memory B lymphocytes into antibody-secreting cells depends on the nature of the thymus-independent carrier used for memory induction and/or revelation, *Cell. Immunol.*, 75, 52, 1983.

130. Werblin, T. P. and Siskind, G. W., Distribution of antibody affinities: technique of measurement, *Immunochemistry*, 9, 987, 1972.

131. Werblin, T. P., Kim, Y. T., Quagliata, F., and Siskind, G. W., Studies on the control of antibody synthesis. III. Changes in heterogeneity of antibody affinity during the course of the immune response, *Immunology*, 24, 477, 1973.

132. Kim, Y. T. and Siskind, G. W., Studies on the control of antibody synthesis. VI. Effect of antigen dose and time after immunization on antibody affinity and heterogeneity in the mouse, *Clin. Exp. Immunol.*, 17, 329, 1974.

133. Gershon, R. K. and Paul, W. E., Effect of thymus derived lymphocytes on amount and affinity of anti-hapten antibody, *J. Immunol.*, 106, 872, 1971.

134. DeKruyff, R. and Siskind, G. W., Studies on the control of antibody synthesis. XIV. Role of T cells in regulating antibody affinity, *Cell. Immunol.*, 47, 134, 1979.

135. Lafrenz, D. E. and Feldbush, T. L., Role of T cells in the development of memory B cells. Quantitative and qualitative analysis, *Immunology*, 44, 177, 1981.
136. Melchers, F. and Andersson, J., B cell activation: three steps and their variations, *Cell*, 37, 715, 1984.
137. Moller, G., Genetic models of B cell differentiation, *Immunol. Rev.*, 64, 1982.
138. Mosier, D. E. and Subbarao, B., T-independent antigens: complexity of B lymphocyte activation revealed, *Immunol. Today*, 3, 217, 1982.
139. Howard, M. and Paul, W. E., Regulation of B cell growth and differentiation by soluble factors, *Ann. Rev. Immunol.*, 1, 307, 1983.
140. Singer, A. and Hodes, R. J., Mechanisms of T cell-B cell interaction, *Ann. Rev. Immunol.*, 1, 211, 1983.
141. Marrack, P., Graham, S. D., Kushnir, E., Leibson, H. J., Roehm, N., and Kappler, J. W., Nonspecific factors in B cell responses, *Immunol. Rev.*, 63, 33, 1982.
142. Falkoff, R. J. M., Zhu, L. P., and Fauci, A. S., Separate signals for human B cell proliferation and differentiation in response to *S. aureus*: evidence for a two-signal model of B cell activation, *J. Immunol.*, 129, 97, 1982.
143. Parker, D. C. and Prakash, S., The effect of culture supernatant from an IL-2 producing cloned T cell hybridoma on anti-immunoglobulin activated B cells, in *B Lymphocytes in the Immune Response*, Klinman, N., Mosier, D. E., Scher, I., and Vitetta, E. S., Eds., Elsevier/North-Holland, Amsterdam, 1981, 193.
144. Jaworski, M. A., Shiozawa, C., and Diener, E., Triggering of affinity enriched B cells: analysis of B cell stimulation by antigen-specific helper factor or LPS. I. Dissection into proliferative and differentiative signals, *J. Exp. Med.*, 155, 248, 1982.
145. Noelle, R. J., Snow, E. C., Uhr, J. W., and Vitetta, E. S., Activation of antigen specific B cells: Role of T cells, cytokines, and antigen in induction of growth and differentiation, *Proc. Natl. Acad. Sci. U.S.A.*, 80, 6628, 1983.
146. Wetzel, G. D. and Kettman, J. R., Mitogen-induced growth of B cells: requirements for the activation and clonal growth from single cells, in *B Lymphocytes in the Immune Response*, Klinman, N., Mosier, D. E., Scher, I., and Vitetta, E. S., Eds., Elsevier/North-Holland, Amsterdam, 1981, 133.
147. Wetzel, G. D. and Kettman, J. R., Activation of murine B lymphocytes. III. Stimulation of B lymphocyte clonal growth with lipopolysaccharide and dextran sulfate, *J. Immunol.*, 126, 723, 1981.
148. Melchers, F., Andersson, J., Lernhardt, W., and Scheier, M. H., H-2 unrestricted polyclonal maturation without replication of small B cells induced by antigen activated T cell help factors, *Eur. J. Immunol.*, 10, 679, 1980.
149. Pure, E., Isakson, P. C., Kappler, K. W., Marrack, P., Krammer, P. H., and Vitetta, E. S., T cell-derived B cell differentiation factors. Dichotomy between responsiveness of B cells from adult and neonatal mice, *J. Exp. Med.*, 157, 800, 1983.
150. Sidman, C. L. and Marshall, J. D., B cell maturation factor: effects on various cell populations, *J. Immunol.*, 132, 845, 1984.
151. DeFranco, A. L., Raveche, E. S., Asofsky, R., and Paul, W. E., Frequency of B lymphocytes responsive to anti-immunoglobulin, *J. Exp. Med.*, 155, 1523, 1982.
152. DeFranco, A. L., Kung, J. T., and Paul, W. E., Regulation of growth and proliferation in B cell subpopulations, *Immunol. Rev.*, 64, 161, 1982.
153. Singer, A., Asano, Y., Shigeta, M., Hathcock, K. S., Ahmed, A., Fathman, C. G., and Hodes, R. J., Distinct B cell subpopulations differ in their genetic requirements for activation by T helper cells, *Immunol. Rev.*, 64, 137, 1982.
154. Ratcliffe, M. J. H. and Julius, M. H., H-2 restricted T-B interactions involved in polyspecific B cells responses mediated by soluble antigen, *Eur. J. Immunol.*, 12, 634, 1982.
155. Brooks, K. H. and Feldbush, T. L., The correlation between the activation state of B cells and their capacity for in vitro propagation of immunologic memory, *Cell. Immunol.*, 76, 213, 1983.
156. Muraguchi, A., Butler, J. L., Kehrl, J. H., and Fauci, A., Differential sensitivity of human B cell subsets to activation signals delivered by anti-u antibody and proliferation signals delivered by a monoclonal B cell growth factor, *J. Exp. Med.*, 157, 530, 1983.
157. Louis, J. A. and Lambert, P.-H., Lipopolysaccharides: from immunostimulation to autoimmunity, *Springer Sem. Immunopathol.*, 2, 215, 1979.
158. Andersson, J., Sjoberg, O., and Möller, G., Mitogens as probes for immunocyte activation and cellular cooperation, *Transplant Rev.*, 11, 131, 1972.
159. Lawton, A. R., Lucivero, G., Levitt, D., and Cooper, M. D., Aspects of mitogen activation of human B cells, in *Human B-Lymphocyte Function: Activation and Immunoregulation*, Fauci, A. S. and Ballieux, R. E., Eds., Raven Press, New York, 1982, 37.
160. Ricci, M., Romagnani, S., Giudizi, G. M., Almerigogna, F., Riagiotti, R., Del Prete, G. F., and Maggi, E., Mechanisms of human B-cell activation by *Staphylococcus aureus*, in *Human B Lymphocyte Function: Activation and Immunoregulation*, Fauci, A. S. and Ballieux, R. E., Eds., Raven Press, New York, 1982, 17.

161. Chiorazzi, N., Fu, S. M., and Kunkel, H. G., Stimulation of human B lymphocytes by antibodies to IgM and IgG. Functional evidence for the expression of IgG on B lymphocyte surface membranes, *Clin. Immunol. Immunopathol.*, 15, 301, 1980.

162. Parker, D. C., Induction and suppression of polyclonal antibody responses by anti-Ig reagents and antigen-nonspecific helper factors: a comparison of the effects of anti-Fab, anti-IgM, and anti-IgD on murine B cells, *Immunol. Rev.*, 52, 115, 1980.

163. Vitetta, E., Pure, E., Isakson, P., Buck, L., and Uhr, J., The activation of murine B cells: the role of surface immunoglobulins, *Immunol. Rev.*, 52, 211, 1980.

164. Howard, M., Mizel, S. B., Lachman, L., Ansel, J., Johnson, B., and Paul, W. E., Role of interleukin 1 in anti-immunoglobulin-induced B cell proliferation, *J. Exp. Med.*, 157, 1529, 1983.

165. Kuritani, T. and Cooper, M. D., Human B-cell differentiation. IV. Effect of monoclonal anti-IgM and anti-IgD antibodies on B cell proliferation and differentiation induced by T cell factors, *J. Immunol.*, 131, 1306, 1983.

166. Muraguchi, A. and Fauci, A. S., Proliferative responses of normal human B lymphocytes. Development of an assay system for human B cell growth factor (BCGF), *J. Immunol.*, 129, 1104, 1982.

167. Ford, R. J., Mehta, S. R., Franzini, A., Montagna, R., Lachman, L. B., and Maizel, A. L., Soluble factor activation of human B lymphocytes, *Nature (London)*, 294, 261, 1981.

168. Cambier, J. C., Monroe, J. G., and Neale, M. J., Definition of conditions that enable antigen-specific activation of the majority of isolated TNP-binding B cells, *J. Exp. Med.*, 156, 1635, 1982.

169. Wetzel, G. D., Swain, S. L., and Dutton, R. W., A monoclonal T cell replacing activity can act directly on B cells to enhance clonal expansion, *J. Exp. Med.*, 156, 306, 1982.

170. Muraguchi, A., Kehrl, J. H., Butler, J. L., and Fauci, A. S., Sequential requirements for cell cycle progression of resting human B cells after activation by anti-immunoglobulin, *J. Immunol.*, 132, 176, 1984.

171. Schimpl, A. and Wecker, W., A third signal in B cell activation given by TRF, *Transplant. Rev.*, 23, 176, 1975.

172. Armerding, D. and Katz, D. H., Activation of T and B lymphocytes in vitro. II. Biological and biochemical properties of an allogeneic effect factor (AEF) active in triggering specific B lymphocytes, *J. Exp. Med.*, 140, 19, 1974.

173. Hoffman, M. K., Macrophages and T cells control distinct phases of B cell differentiation in the humoral immune response in vitro, *J. Immunol.*, 125, 2076, 1980.

174. Hunig, T., Schimpl, A., and Wecker, E., Mechanism of T cell help in the immune response to soluble protein antigens. I. Evidence for in situ generation and action of T cell replacing factor during the anamnestic response to DNP-KLH in vitro, *J. Exp. Med.*, 145, 1216, 1977.

175. Hunig, T., Schimpl, A., and Wecker, E., Mechanism of T cell help in the immune response to soluble protein antigens. II. Reconstitution of primary and secondary in vitro immune responses to DNP-carrier conjugates by T cell replacing factor, *J. Exp. Med.*, 145, 1228, 1977.

176. Endres, R. O., Kushnir, E., Kappler, J. W., Marrack, P., and Kinsky, S. C., A requirement for nonspecific T cell factors in antibody responses to "T cell independent" antigens, *J. Immunol.*, 130, 781, 1983.

177. Leibson, H. J., Marrack, P., and Kappler, J., B cell helper factors. II. Synergy among three helper factors in the response of T cell and macrophage-depleted B cells, *J. Immunol.*, 129, 1398, 1982.

178. Nakanishi, K., Howard, M., Muraguchi, A., Farrar, J., Takatsu, K., Hamaoka, T., and Paul, W. E., Soluble factors involved in B cell differentiation: identification of two distinct T cell replacing factors, *J. Immunol.*, 130, 2219, 1983.

179. Pure, E., Isakson, P. C., Takatsu, K., Hamaoka, T., Swain, S. L., Dutton, R. W., Dennert, G., Uhr, J. W., and Vitetta, E. S., Induction of B cell differentiation by T cell factors. I. Stimulation of IgM secretion by products of a T cell hybridoma and a T cell line, *J. Immunol.*, 127, 1953, 1981.

180. Isakson, P. C., Pure, E., Vitetta, E. S., and Krammer, P. H., T cell derived B cell differentiation factor(s). Effect on the isotype switch of murine B cells, *J. Exp. Med.*, 155, 734, 1982.

181. Layton, J. E., Vitetta, E. S., Uhr, J. W., and Krammer, P. H., Clonal analysis of B cells induced to secrete IgG by T cell-derived lymphokine(s), *J. Exp. Med.*, 160, 1850, 1984.

182. Kawanishi, H., Saltzman, L., and Strober, W., Mechanisms regulating IgA class specific immunoglobulin production in murine gut-associated lymphoid tissues. II. Terminal differentiation of post-switch sIgA-bearing Peyer's patch B cells, *J. Exp. Med.*, 158, 649, 1983.

183. Cebra, J. J., Komisar, J. L., and Schweitzer, P. A., C_H isotype "switching" during normal B lymphocyte development, *Ann. Rev. Immunol.*, 2, 493, 1984.

184. Zubler, R. H. and Glasebrook, A. L., Requirements for three signals in "T-independent" (LPS induced) as well as in T-dependent B cell responses, *J. Exp. Med.*, 155, 666, 1982.

185. Tittle, T. V. and Rittenberg, M. B., IgG B memory cell subpopulations: differences in susceptibility to stimulation by TI-1 and TI-2 antigens, *J. Immunol.*, 124, 202, 1980.

186. Feldbush, T. L., Unpublished observation, 1984.

187. Snow, E. C., Noelle, R. J. Uhr, J. W., and Vitetta, E. S., Activation of antigen-enriched B cells. II. Role of linked recognition in B cell proliferation to TD antigens, *J. Immunol.*, 130, 614, 1983.

188. Shiozawa, C., Sawada, S., Inazawa, M., and Diener, E., Triggering of affinity enriched B cells. II. Composition and B cell triggering properties of affinity purified antigen specific, T cell derived helper factor, *J. Immunol.*, 132, 1892, 1984.

189. Pike, B. L., Vaux, D. L., and Nossal, G. J. V., Single cell studies on hapten-specific B lymphocytes: differential cloning efficiency of cells of various sizes, *J. Immunol.*, 131, 554, 1983.

190. Pike, B. L., Vaux, D., Clark-Lewis, I., Schrader, J. W., and Nossal, G. J. V., Proliferation and differentiation of single hapten-specific B lymphocytes promoted by T cell factor(s) distinct from T cell growth factor, *Proc. Natl. Acad. Sci. U.S.A.*, 79, 6350, 1982.

191. Snow, E. C., Vitetta, E. S., and Uhr, J. W., Activation of antigen-enriched B cells. I. Purification and response to TI antigens, *J. Immunol.*, 130, 607, 1983.

192. Andersson, J. and Melchers, F., T cell dependent activation of resting B cells: requirement for both nonspecific unrestricted and antigen specific Ia-restricted soluble factors, *Proc. Natl. Acad. Sci. U.S.A.*, 78, 2497, 1981.

193. Zlotnik, A., Roberts, W. K., Vasil, A., Blumenthal, E., Larosa, F., Leibson, H. J., Endres, R. O., Graham, S. D., White, J., Hill, J., Henson, P., Klein, J. R., Bevan, M. J., Marrack, P., and Kappler, J. W., Coordinate production by a T cell hybridoma of gamma interferon and three other lymphokine activities: multiple activities of a single lymphokine? *J. Immunol.*, 131, 794, 1983.

194. Tada, T., Takemori, T., Okumura, K., Nonaka, M., and Tokuhisa, T., Two distinct types of T helper cells involved in the secondary antibody response: independent and synergistic effects of Ia− and Ia+ T_H cells, *J. Exp. Med.*, 147, 446, 1978.

195. Takatsu, K., Sans, Y., Hashimoto, N., Tomita, S., and Hamaoka, T., Acceptor sites for T cell replacing factor (TRF) on B lymphocytes. I. TRF-substituting activity of anti-TRF acceptor site(s) antibody in triggering of B cells, *J. Immunol.*, 128, 2575, 1982.

196. Tominaga, A., Takatsu, K., and Hamaoka, T., Antigen-induced TRF. II. X-linked gene control for the expression of TRF acceptor sites on B cells, *J. Immunol.*, 128, 2581, 1982.

197. Takatsu, K., Tanaka, K., Tominaga, A., Kumahara, Y., and Hamaoka, T., Antigen-induced T cell replacing factor (TRF). III. Establishment of T cell hybrid clone continuously producing TRF and functional analysis of released TRF, *J. Immunol.*, 125, 2646, 1980.

198. Julius, M. H., Chiller, J. M., and Sidman, C. L., Major histocompatibility complex-restricted cellular interactions determining B cell activation, *Eur. J. Immunol.*, 12, 627, 1982.

199. Hirano, T., Teranishi, T., Lin, B., and Onoue, K., Human helper T cell factor(s). IV. Demonstration of human late acting BCDF acting on S. aureus Cowan I-stimulated cells, *J. Immunol.*, 133, 798, 1984.

200. Rennick, D. M., Morrow, P. R., and Benjamini, E., Two stages of B cell memory development expressing differential sensitivity to stimulation with TD and TI antigenic forms, *J. Immunol.*, 131, 567, 1983.

201. Rennick, D. M., Morrow, P. R., and Benjamini, E., Functional heterogeneity of memory B lymphocytes: in vivo analysis of TD-primed B cells responsive to secondary stimulation with TD and TI antigens, *J. Immunol.*, 131, 561, 1983.

202. Brooks, K. H. and Feldbush, T. L., In vitro antigen-mediated clonal expansion of memory B lymphocytes, *J. Immunol.*, 127, 959, 1981.

203. Howard, M. C., Baker, J. A., and Shortman, K., Antigen-initiated B lymphocyte differentiation. XV. Existence of "preprogenitor" and "direct progenitor" subsets among secondary B cells, *J. Immunol.*, 121, 2066, 1978.

204. Thomas, K. R. and Feldbush, T. L., Manuscript in preparation.

Interleukin-2 receptor, 43, 48—49
Interleukin-X, 64—67, 151
Intracellular transit time, 36
In vitro nuclear pulse labeling, 31
In vivo-in vitro differences in B cell activation, 48
In vivo pulse labeling, 31
Isotype switching, 141

J

J chain genes, 24, 27, 29, 33—34, 36

K

K46R, 27
Kappa chain gene rearrangement, 118
Keyhole limpet hemocyanin (KLH), 62
Kupffer cells, 120

L

L10A, 27
L10A/2J tumors, 121
Langerhans' cells, 120
Large B cells, 90
Late-acting B cell helper, 72
L chain genes, 24
Ligand receptor interactions, 3
Light chain gene rearrangement, 112, 117
Light-chain genes, 29
Light chain mRNA, 114
Linked recognition, 68, 151
Lipopolysaccharide (LPS), 102, 112, 118, 120—121, 147—149
  mitogenic stimulation of B cells, 7
  stimulation, 24—25, 29, 31, 33—37, 43, 49
Liquid microculture, 82
Low epitope density antigens, 51—52
Ly-1 B cell, 116—117
Lyb-2 antigen, 115
Lyb 5$^+$ B cells, 97, 145, 149
Lyb 5$^-$ B cells, 97
Ly-m19 antigen, 115
Lymphoblasts, 3
Lymphocyte activating factor (LAF), 66
Lymphocytes
  precursor frequency, 154
  recirculation, 136
Lymphoid malignancies, 113
Lymphokine cascade, 88, 92
Lymphokine dependence of TI antigens, differentiation of single B cells, 91—97
Lymphokine-independent antigens, 94
Lymphokines, 45, 50, 81, 86, 102
  cloned, 72
  effects of, 92—93
Lymphoma cells, 32—33
Lymphomas, 27, 113, 116

1,29,115

M

Macromolecular metabolism, 5
Macrophage factors, 150
Macrophages, 120
Major histocompatibility complex (MHC), 62, 68, 81, 120
Malignant B cells, 114—126
  antigen presentation, 119—120
  antigen-specific, 122—123
  CH5, 116
  differentiation antigens, 114—117
  induced differentiation, 117—126
  Ly-1 B cell subpopulation, 116—117
  mature, 118—125
  monoclonal antibody preparation, 114—116
  polyclonal activation, 120—122
  pre-B cell tumors, see Pre-B cell tumors
  surface immunoglobulin role in B cell activation, 123—124
Mature B cell malignancies, 118—125
Mature B cells, 92
Mel 14, 43
Membrane depolarization, 2—3, 7—9, 12, 14
  calcium ionophore A23187, 9, 11
  mIa expression, 7
  phorbol myristate acetate, 9
  protein kinase C, 7—9
Membrane immunoglobulins, see mIg
Memory B cell, see Memory cell
Memory B lymphocytes, see Memory cell
Memory cell, 25, 80, 112, 135—162
  antigen-specific activation, 150—152
  expansion requirements, 152—154
  generation, 145—147
  historical perspective, 136—137
  immature, 137—139
  immature and mature, 136
  immunoglobulin class commitment, 139—143
  immunoglobulin secretion, 141
  induction of memory, 143—147
  large, 137—138, 140
  maturation process, 137
  mature, 136, 139
  mature and immature, 136
  polyclonal activation, 147—150
  precursor population, 143—145
  small, 138—140
  specific isotypes, production of, 141
  subsets, 137—143
  surface isotype, 139—143
  T cell role in B memory generation, 136, 145—147
Methylation, 3
Methylation status of expressed genes, 26—28
mIa expression, 7
mIg, 2, 25, 80
  cross-linking, 2, 7—9, 11—16, 25

S

T

U

UNM lymphomas, 118—119

V

Valinomycin, 7
VDJ joining, 27, 36
Verapamil, 6
Virgin B cells
  antigen-specific activation, 150—152
  expansion requirements, 152—154

polyclonal activation, 147—150

W

WEHI lymphomas, 118
WEHI-231 lymphomas, 24, 27, 116, 120, 122

X

Xid immune deffect, 48
X-linked immunodeficiency syndrome, 92

Printed and bound by CPI Group (UK) Ltd, Croydon, CR0 4YY

22/10/2024

01777632-0018